Artefact Kinds

SYNTHESE LIBRARY

STUDIES IN EPISTEMOLOGY, LOGIC, METHODOLOGY, AND PHILOSOPHY OF SCIENCE

VOLUME 365

For further volumes:
http://www.springer.com/series/6607

Maarten Franssen • Peter Kroes
Thomas A.C. Reydon • Pieter E. Vermaas
Editors

Artefact Kinds

Ontology and the Human-Made World

Springer

Editors
Maarten Franssen
Section of Philosophy
Delft University of Technology
Delft, The Netherlands

Peter Kroes
Section of Philosophy
Delft University of Technology
Delft, The Netherlands

Thomas A.C. Reydon
Institut für Philosophie
Leibniz Universität Hannover
Hannover, Germany

Pieter E. Vermaas
Section of Philosophy
Delft University of Technology
Delft, The Netherlands

ISBN 978-3-319-00800-4 ISBN 978-3-319-00801-1 (eBook)
DOI 10.1007/978-3-319-00801-1
Springer Cham Heidelberg New York Dordrecht London

Library of Congress Control Number: 2013948358

Printed on acid-free paper

Springer is part of Springer Science+Business Media (www.springer.com)

Contents

Part III Artefact Kinds and Engineering Practice

Chapter 1
Introduction: The Ontology of Technical Artefacts

Maarten Franssen, Peter Kroes, Thomas A.C. Reydon, and Pieter E. Vermaas

Abstract This chapter introduces the reader to the main difficulties that artefacts pose for ontology. Due to the possibilities of disassembly and reassembly, it is problematic to come up with clear identity conditions for artefacts. Due to the possibility of radical reassembly of parts to create an artefact of another kind, artefact classification is problematic when interpreted ontologically. Since the (re) assembly of artefact components is done by people with certain purposes in mind, artefact ontology seems to depend crucially, and from a metaphysical point of view problematically, on the mental states of humans. The chapter then summarizes how the various contributions in the book discuss aspects of these difficulties and explains how contributions that take an epistemological or ethnographic perspective show the common ground between the contributions that address the problem of artefact kinds from a traditional philosophical perspective and the contributions that focus on artefact classification and "ontology engineering" from an engineering perspective.

Keywords Artefacts • Artefact kinds • Classification • Engineering • Ontology • Technology

---

M. Franssen (✉) • P. Kroes • P.E. Vermaas
Section of Philosophy, Delft University of Technology,
Jaffalaan 5, 2628 BX Delft, The Netherlands
e-mail: m.p.m@tudelft.nl; p.a.kroes@tudelft.nl; p.e.vermaas@tudelft.nl

T.A.C. Reydon
Institut für Philosophie, Leibniz Universität Hannover,
Im Moore 21, 30167 Hannover, Germany
e-mail: reydon@ww.uni-hannover.de

M. Franssen et al. (eds.), *Artefact Kinds: Ontology and the Human-Made World*,
Synthese Library 365, DOI 10.1007/978-3-319-00801-1_1,

1.1 What Is the Problem? Artefact Identity, Artefact Classification and Artefact Kinds

This volume is concerned with two intimately related topics that occupy central positions in metaphysics: the identity of entities and the foundations of classification. Both topics are of interest for theoretical as well as practical reasons and, accordingly, have drawn a considerable amount of attention not only from metaphysicians but also from researchers working in other areas of philosophy, such as philosophy of science (and within that domain most prominently philosophy of biology), philosophy of language and – more recently – practice-oriented ontology (formal ontology, bio-ontology and the like).

Most of the philosophical work done so far on identity and classification has addressed the identity and classification of naturally occurring entities – atoms, substances, animals, human beings, etc. The present volume, however, addresses these issues with respect to a different category of things. Here, the focus lies on the identity and classification of artefacts, because – as the following chapters show – for artefacts these questions require different answers than they do in the case of natural entities. In this respect, the identity and classification of artefacts is a problem domain that should be of interest to philosophers both for its own sake and because the solutions proposed there might shed new light on the corresponding issues for the case of natural entities.

For a long time throughout the history of philosophy, artefacts have been a neglected domain. Artefacts came into focus as objects of philosophical study only at the beginning of the modern era, which may be connected to the declining influence of Aristotle at that time. (As is well known, Aristotle granted artefacts only a secondary place in his ontology.) In this respect, Thomas Hobbes is a key figure.[1] Hobbes' famous reference to the ship of Theseus, which he made in the context of his discussion of the principle of individuation and to which we shall turn shortly, marks the onset of a line of philosophical thinking that continues to the present day.

While in contemporary philosophy there is a clear interest in artefacts (e.g. Margolis and Laurence 2007), still artefacts are often held in low esteem by philosophers, as things somehow not participating fully in the architecture of things. For example, in her recent book on *The Metaphysics of Everyday Life* (2007), Lynne Rudder Baker presents

[1] Hobbes at one time was personal secretary to Francis Bacon. Bacon famously opposed the logic of the Aristotelian *Organon*, which was still the dominant method of reasoning at the time, and proposed a new logic as more suitable for the investigation of nature in his *New Organon* (published in 1620 as part of his "great renewal of the sciences", the *Instauratio Magna*). While this made Bacon one of the initiators of the Scientific Revolution, one of Bacon's principal concerns was with artefacts. In the *New Organon*, he repeatedly emphasised that the investigation of nature should not primarily be seen as a goal in itself, but should stand in the service of constructing new kinds of artefacts that could be used to improve the living circumstances of the people. This focus on artefacts also plays a prominent role in Bacon's utopian work, the *New Atlantis* which was written in 1610 but published only after Bacon's death.

the following list of criteria for entities to be what she calls "genuine substances" – or entities that are "irreducibly real" – and suggests that artefacts have been discredited on all of them (p. 60):

1. Fs are genuine substances only if Fs have an internal principle of activity.
2. Fs are genuine substances only if there are laws that apply to Fs as such or there could be a science of Fs.
3. Fs are genuine substances only if whether something is an F is not determined merely by an entity's satisfying some description.
4. Fs are genuine substances only if Fs have an underlying intrinsic essence.
5. Fs are genuine substances only if the identity and persistence of Fs are independent of any intentional activity.

These five criteria can serve as a summary for the reasons why artefacts often are not taken seriously as full members of the furniture of the world and, by consequence, are not seen as being of interest to metaphysicians *as artefacts*. (To be sure, metaphysicians regularly mention artefacts, but when they do they often merely use them as examples of "medium-sized dry goods" in discussions on broader metaphysical topics – they are not often studied qua artefacts.) The first four criteria hang closely together, and as such, as will be made clear below, represent an influential view on what is "real" in the structure of reality, but not one that straightforwardly entails that artefacts are not genuine entities. The fifth criterion stands on its own and seems to discredit artefacts without the possibility of an appeal, since, as the definitions of Risto Hilpinen (1992, 1993) and Randall Dipert (1993) indeed have it, artefacts are, one way or another, the product of human intentional activity. The fifth criterion, however, may well be far too strong as it stands; would it not rule out either the genuineness of human beings or a role for human intentionality in the persistence of human beings, against what we take to be the case?

But let us return to earlier philosophers to see why artefacts qua artefacts should be of interest to the metaphysician. It is from Plutarch and a number of earlier authors that we know the story of the ship of Theseus: The Athenians wanted to keep the ship that Theseus used to sail to Crete and kill the Minotaur in their harbour. They preserved the ship by gradually replacing the planks that had deteriorated too much by new ones. But already in antiquity this practice led to a discussion whether the preserved ship, which ultimately consisted mostly of new planks, was still the same ship as Theseus' original one.

Hobbes mentioned the ship of Theseus in his *De Corpore* of 1655, in considering the problem that although "one and the same thing can be compared with itself, although only at different times" it is not clear "in what sense should a body sometimes be deemed to be the same body, and sometimes a different one from what it was before".[2] In other words, what is unclear is which principle of individuation comes into play when identifying a later entity and an earlier entity as the same

[2] Part 2, Chapter 11, Section 7. Citations are from the anonymous English translation published in 1656, which Hobbes authorised.

entity. However, Hobbes added a twist to the story, suggesting that someone could have hoarded the removed planks and built a second ship out of them, a ship with an equal if not better claim to be the ship of Theseus as the ship the Athenians wanted to preserve. This allowed him to pit two principles of individuation against one another: the principle that unity of matter (consisting of the same material components) is what makes two objects at different times the same object and the principle that the unity of form (presumably, conforming to the same construction plan) is the relevant factor. Hobbes argued against unity of form being the single principle of individuation, because this would force us to identify *two* later ships with one earlier ship as being the ship of Theseus. Neither can unity of matter be the exclusive criterion of identity, since it would rule out the possibility of identity over time for metabolising organisms, including humans. Hobbes concluded that "the beginning [*principio*] of individuation is not always to be found either from matter alone, or from form alone".

What is at stake in both the original story and Hobbes' version is the identity of a particular entity: what, exactly, determines the identity of this particular ship that we have in front of us right now as (not) being Theseus' ship? Hobbes' version of the story, however, brings to the fore an important aspect that was covered up in the original story. In the original story, the problem was what determines the identity of an entity qua particular entity while it undergoes profound changes in its material constitution. The question here seemed to be whether historical continuity – that is, the ship being traceable back in time to Theseus' ship through a continuous line of ships at different times – is a *sufficient* condition for determining the ship's identity. Perhaps not: perhaps historical continuity in the sense of a continuous line of ships in time is a *necessary* condition, and being made up of the original material is required too, or perhaps historical continuity is neither a sufficient nor a necessary condition for the ship's identity.

By suggesting that the ship's identity could be preserved through disintegration and reassembly, Hobbes pointed to two additional elements of the identity problem. The first element is that the problem manifests itself differently for artefacts (which can be disassembled and reassembled) than for natural entities (where disintegration means ceasing to exist). The second element is that kind membership – where unity of form is an essential element of kind membership – is often an important factor in identity questions. Hobbes took it to be possible that an entity (Theseus' ship) disintegrates and thus ceases to exist, but at a later time *the same entity* comes into being again. Complicating the story even more, all the removed planks could be placed on a large heap before the second ship is built, or the removed planks could be used to build a boat shed, which is later torn down again, the planks being then used to build another ship. In the time window between the old ship and the new ship, there exists a thing (the heap or the boat shed) that is materially continuous with but not identical to either of the two ships: the thing that exists in the intermediate time window is of the wrong kind (it has the wrong form). By considering the case of the removed planks being hoarded, Hobbes' version of the story thus points out why artefacts should be of metaphysical interest for their own sake (as there are identity issues that arise mainly if not only for artefacts). Additionally, it emphasises that the metaphysics of identity is intimately connected to the metaphysics of classification, a topic that was later picked up by John Locke in his highly influential *Essay Concerning Human Understanding* of 1690.[3] Where the two issues meet is in the metaphysics of *kinds*.

Metaphysical problems involving the existence and identity of artefacts in particular are related to, but should be distinguished from, metaphysical problems about the existence of individual objects in general. The latter problems have to do with the unity that such objects have with respect to the underlying matter. An example of these problems is presented by the case of a statue made out of some material, say clay.[4] It seems we can distinguish between two objects, one the lump of clay, the other the statue made out of it. The lump of clay existed before a statue was made out of it and will continue to exist once the statue is destroyed by squeezing it out of existence. As long as we accept either as an entity, the lump and the statue must, therefore, be distinct entities, but once we have acknowledged this, we must accept that two distinct entities simultaneously occupy precisely the same region of space, a conclusion that many find counterintuitive. To avoid conclusions like these, some philosophers have maintained that no individual objects exist at all; what exists is matter occupying regions of space (e.g. Jubien 1993 and, more implicitly, Goodman and Leonard 1940). Familiar objects are mere temporal phases in the existence of these chunks of matter. In such a sparse metaphysics, no artefacts exist because no ordinary objects exist, in the sense of being to a large extent independent of the precise matter from which they are composed. Other philosophers have arrived at equally sceptical conclusions cued by the scientific picture that what we take to be ordinary objects are actually highly complex arrangements of smaller particles. They argue that no unambiguous identifications of ordinary objects with particular arrangements are possible or that ordinary objects fall out of the causally closed universe described by physics (e.g. Unger 1979; van Inwagen 1990; Merrick 2001). Familiar objects are mere epiphenomena with respect to the existence of simple particles.[5]

The problem of the metaphysical status of artefacts in particular, therefore, poses itself only if the cogency of the existence of individual objects is acknowledged: objects whose unity consists only partly in their composing matter but additionally, or even primarily, in some constitutive principle, which makes this unity to some extent independent of the composing matter. Such individual objects can gain or lose parts and remain the object that they are. Those who accept the existence of individual objects can point out paradigmatic examples everywhere in nature: animals, plants, gems, rivers and so forth. An individual tiger metabolises and accordingly is in constant flux at the molecular level. It survives losing its hairs, its

[3] Locke's *Essay*, of course, also opposed the Aristotelian/Scholastic way of thought. See for Locke's metaphysics also (Ayers 1991), in particular ch. 21 on 'Artificial and other problematical objects'.

[4] The example is due to Allan Gibbard (1975).

[5] Notoriously, van Inwagen excludes living beings from his scepticism, and Merrick excludes persons. For more on these issues, see, e.g., Rea (1996).

whiskers, its claws, its teeth and, occasionally, its tail or an entire paw. It is on this unity, which can arise in particular quantities of matter and then is maintained with some degree of independence of this composing matter, that the metaphysical status of artefacts turns. The unity of an individual animal is something that seems to us to be given. The principles that keep it coherent and intact are principles based on, or somehow wrought from, nature. It is for us to discover how strong and cohesive natural entities are, that is, what happens in and to a tiger throughout changes in the tiger and in its environment; what properties a tiger can lose and still survive and what will put an end to a tiger. What is more, the beginnings and ends of tigers are well-defined phenomena, which we cannot change the status of: we cannot keep a tiger in existence beyond its death. All of these things seem exactly false for artefacts, as the case of the ship of Theseus makes abundantly clear. The possibility of disassembly and reassembly lends the going out of existence of an artefact a high degree of arbitrariness. If a ship is disassembled, the planks stored for some time and then put together again, we find it hardly controversial to say this is the same ship. But what if for some reason the planks are never reassembled? Did the ship go out of existence when it was taken apart or when it became clear that it would never be reassembled? In the case of the ship of Theseus, moreover, the disassembly, replacement and reassembly takes place in such a way that there are two candidates for being the original ship, again introducing a degree of arbitrariness in the issue of the ship's continued existence. In fact, the ship's continued existence appears to depend entirely on how we choose it. This dependence on human minds seems unacceptable for an item's "real" existence: if it exists, it should exist independently of what we care to think about it.

This kind of mind-dependence should be sharply distinguished from the historic mind-dependence of artefacts, the fact that they come into being as a result of intentional human action. It is not so clear that this aspect would make any artefact metaphysically suspect.[6] After all, many people also owe their coming into being to an intentional decision of minimally one person to set into motion the fairly reliable procedure that eventually results in the birth of a child. So the metaphysically crucial dependence on human intentions is not so much the historic one that goes into the definition of any artefact, as the dependence on intentions for the question whether an artefact is still there or is still the same artefact as an earlier identified one. David Wiggins emphasised this by even *defining* "artefact" as an object that has some unity with respect to its composing matter but lacks a clear principle of unity or organisation dictated by the laws of nature, such as is had by the paradigmatic natural kinds of things mentioned previously (1980, p. 89.). This definition does not take into account the object's origin at all, and it excludes, for example, intentionally made instances of natural kinds, such as synthetic diamonds, from being artefacts but includes in the category of artefacts things found in nature like wasps' nests and spiders' webs. The seeming dependence of artefacts on mind and intentions also includes a dependence of the *kind* of artefact a thing is on intentions,

[6] See on this issue esp. Thomasson (2007).

since changes in the perception of an artefact or the use it is put to seemingly may result in an artefact "becoming another thing".

Indeed the metaphysical status of artefacts compared to natural objects is closely associated with the kind membership and classification of things, and the questionable metaphysical status of artefacts shows itself here in various ways. If individual objects are identifiable as such through a unifying and organising principle based upon the laws of nature and acting on some amount of matter, then it is plausible that objects also are identifiable as objects of a particular kind, the kind being determined by the repeatable operation of the unifying and organising principle in different amounts of matter. So much seems assured by the lawlike connections underlying the principle, which are secured by the lawlike behaviour of matter. Any principle of identity requires some constancy or regularity of behaviour in order to be applicable. Classification is the grouping of individual things into larger units, linked by certain regularities and similarities. For natural kinds of things, classification is difficult to separate from identification in the first place. A particular object could be identified and thus classified as an animal, or a mammal, or a tiger; the difference seems largely one of accuracy. Classification, moreover, seems basically a matter of discovery, for example, when it was discovered that whales are mammals, not fish. As a result, not just the individual kinds but also the various hierarchical relations between them become included in the "real" ontological structure of the world. All this seems lacking for artefacts. Artefacts, just as much as natural objects, can be identified at arbitrary levels of generality, say, as a ship, or as a sailing boat, or as a item of transportation, or as a Transpac 52 Class yacht, yet it seems impossible to come up with a hierarchy of organisational principles governed by the laws of nature that applies to all these classes. It is exactly at the general level of "ship" or "clock" that people despair of formulating unifying or organising principles that serve to identify any ship or clock as an instance of a more general class. Moreover, what goes into a ship or clock – any ship or any clock – seems not a matter of discovery but of human ingenuity. Whether or not something counts as a clock or a ship will often be settled by convention, which brings us back to the mind-dependence of artefact categories. The diversity among artefact categories seems much larger than among natural kinds of things, however, and the difficulties surrounding artefact classification may not be equally severe at all levels. How the practice of classification works in technology and what the details of this practice imply for the metaphysical status of artefacts and artefact kinds is a matter that has not yet received much philosophical attention.

Notwithstanding the many arguments intended to show that artefacts and artefact kinds cannot have a metaphysical status that is as strong as the status of natural things and natural kinds, this conclusion seems highly problematic and counterintuitive. In our daily lives, artefacts play roles comparable to the roles that instances of natural kinds – people, animals, plants, geological items – play, and tables, can openers, vacuum cleaners, cars and the like do not seem less real to us in any way. Philosophers interested in technology therefore are interested in investigating the arguments for a sharp metaphysical divide separating natural kinds of things from artefactual kinds. The suggestion that artefacts and artefact kinds can be salvaged as objects and kinds just as metaphysically real as natural objects and natural kinds of things if only a particular level of detail is chosen is one way in which a solution has been sought. Other approaches question the sharp separation between artefacts and natural objects and argue that natural kinds of things suffer from vagueness and arbitrariness in much the same way as artefact kinds do. The clarity of the unifying principles for natural objects may therefore also be questioned. Finally the character of the mind-dependence of the unifying principles for artefacts may be further specified; as stated, a purely historical form of mind-dependence arguably is metaphysically innocuous, but the same could perhaps be argued for nonhistorical forms.

1.2 The Structure and Content of the Book

The above considerations make clear that discussions on the character of artefacts and artefact kinds have a bearing on important ontological and epistemological questions central to philosophy. In the contributions to this volume, many of these questions will be touched upon in the context of the specific problems discussed by the various authors. These discussions focus primarily on technical artefacts, material objects that are made and/or adopted to be made use of. Social artefacts, understood as a general term for more or less abstract arrangements and constructs but with an instrumental role to play, similar to technical artefacts, are only occasionally dealt with. Problems specific to works of art of whatever nature, for which it is doubtful whether they can be understood in relation to a context of *use* at all, fall outside the scope of this volume.

The volume is structured into three parts, beginning with the ontology of artefact kinds, then moving on to meta-issues with regard to the ontology of artefact kinds and ending at the more concrete level of ontologies of artefact kinds in engineering practice. The following is an overview of the chapters included in this volume.

The contributions in Part I address basic ontological and metaphysical questions that arise in relation to artefact kinds: What *are* artefact kinds? Are they real kinds? And what makes individual artefacts members of their kinds? That is, how should we conceive of the nature of artefact kinds and how are artefact kinds to be defined? What are the identity conditions for individual artefacts and artefact kinds? Although the authors start off more or less from the same question, what artefact kinds are, they provide quite different answers to this question. One of the main bones of contention is whether technical artefact kinds are mind-dependent or not.

In Chap. 2, Jonathan Lowe argues that the nature of artefact kinds is to be understood in terms of law-based modes of activity. For many artefact kinds no clear conditions of existence and, in particular, persistence can be stated for the things belonging to that kind. The existence and persistence conditions for natural kinds are grounded in natural laws, Lowe argues, but no similar laws seem to exist for artefacts. Lowe argues that this may be so for the subcategory of utensils (chairs, hammers and the like) but not for the subcategory of machines: these are governed by sortal-specific engineering laws capable of doing the metaphysical grounding work for machines that natural laws do for natural kinds. This, Lowe concludes, should lead us to think of the category of artefacts as a heterogeneous category: some artefact kinds (kinds of utensils) cannot be thought of as real kinds, because no nonarbitrary existence and persistence conditions can be identified for their member entities. Other kinds of artefact kinds (kinds of machines), however, *can* be thought of as real kinds, because nonarbitrary law-based existence and persistence conditions for their member entities can be formulated. Because these real artefact kinds are grounded in laws, they are mind-independent.

In Chap. 3, Crawford Elder also defends a position according to which artefact kinds are mind-independent. He argues for a view in which artefact kinds are defined by means of proper functions. He focuses on the causal connections that secure the clustering of the essential properties of real kinds. For artefact kinds, he argues, Millikan's etiological theory of proper function can furnish the causal securement of the clustering of morphology, function and historical proper placement, by which artefact kinds can be individuated. As a consequence of this view, an artefact's function is what in fact caused it to be reproduced, which need not be identical to what it was designed for. Only in this way can the mind-dependence of artefacts and artefact kinds be successfully avoided. Finally, by defusing the argument that facts about cultural artefacts, such as gender roles and codes of etiquette, are made true because "thinking makes it so", Elder concludes that even cultural "artefacts", which have no morphology to speak of, can be saved for realism.

Amie Thomasson, in Chap. 4, by contrast proposes to conceive of artefact kinds as being defined by the features of artefacts that were intended by their creators, thus emphasising intentional rather than causal relations in characterising artefacts. She argues that for understanding the role of public artefacts in our daily lives, artefacts are better conceived of as objects with certain intended features and not, as is common, as objects intentionally made to serve a purpose, that is, as objects with a certain function. Beyond functional, structural and perceptible properties, these intentional features may include receptive and normative features that relate to how created objects are to be regarded, used, treated, etc. She claims that these receptive and normative features are typically among the essential features of members of public artefact kinds. Apart from solving issues about exaptation and minimal creation, this broader conception of artefacts may help in understanding their significance for and normative role in daily life.

Next, in Chap. 5, Maarten Franssen and Peter Kroes defend a view in which artefact kinds are defined by means of their structure in combination with their history. They distinguish between the metaphysical discrediting of artefacts as contained in the discrediting of ordinary objects in general and the metaphysical discrediting of artefacts on the basis of criteria that address artefacts in particular. A discussion of recent defences of the metaphysical reality of ordinary objects reveals that there are additional criteria at play in discrediting certain kinds as only nominal than just the criterion of mind-dependence or indeterminateness discussed with respect to artefacts in particular. Using the conjunction of these criteria, Franssen and Kroes argue that artefact kinds can be defended as real in the form of historical sub-kinds of structural kinds.

In the concluding Chap. 6 of Part I, Massimiliano Carrara, Silvia Gaio and Marzia Soavi examine formal constraints on identity criteria for artefacts and show how the well-known problem with transitivity of identity criteria for artefacts may be resolved in order to avoid an alleged problem regarding the ontological respectability of artefacts. They take up the problem of artefact persistence through changes over time, an issue that also played a central role in Lowe's discussion in Chap. 2. One of the reasons for judging artefact kinds not to be real kinds is that their identity conditions fail to satisfy the formal constraint of transitivity. In particular the two prime candidates for relations grounding artefact identity, sameness of parts and sameness of function, are not equivalence relations. They show, however, that a logically adequate approximating relation, satisfied or at least satisfiable by artefacts, can take the place of the required equivalence relation.

The contributions included in Part II address what may be called meta-ontological questions. In focus here are questions that logically precede attempts at developing an ontology of artefact kinds; these are questions such as: What, exactly, should an ontological account of artefact kinds provide us with? Which ways of approaching the ontology of artefact kinds can we choose from? Which approaches might be more promising than others and why? What should the scope of an ontology of artefact kinds be and what are the criteria by which to assess an ontology of artefact kinds?

In Chap. 7 Jesús Vega Encabo and Diego Lawler analyse the challenge that the appearance of new kinds of artefacts in the world poses for ontological theories of artefact kinds. New kinds of artefacts are designed and made all the time, they point out. Many are variations upon already existing themes, but some are genuinely novel – occasionally completely new kinds of things are created, things that the world has not seen yet. Therefore, any adequate metaphysical account of artefact kinds should meet what Vega and Lawler call the *creation requirement*: any theory of artefact kinds must be able to account for the identity of genuinely novel artefacts (which, by being genuinely novel, do not belong to any existing artefact kind), and, moreover, it must be able to explicate not only the ontology of existing artefact kinds but also their ontology with respect to how they come into being. Vega and Lawler take a more focused perspective on technical artefacts and consider the creation of technical artefacts of genuinely new kinds from an ontological point of view. They argue that a combination of functional and intentional theories of the nature of artefacts is required to make sense of artefacts and artefact kinds, and they describe aspects that an ontological theory should take into account in order to make room for new artefact kinds.

In Chap. 8, Thomas Reydon turns to the meta-question how we might try to get to a theory of artefact kinds. His answer is that we first should look at the epistemic roles of artefact kinds in the academic engineering disciplines and in professional design and connect that enterprise to what is done in the philosophy of natural kinds before we turn to the metaphysics. He compares two different ways in which one might attempt to develop a philosophical theory of artefact kinds, namely, approaches that start from metaphysical considerations and approaches that address the epistemology of artefact kinds first and place metaphysics second in line. According to Reydon, discussions about the metaphysical status of artefacts and artefact kinds suffer not only from a lack of consensus about which objects are considered to belong to the domain of artefacts and which not, but also about the metaphysical criteria for kindhood in general. Inspired by recent discussions about natural kinds in the philosophy of biology, he argues that a way out of this situation may be found in a so-called epistemological turn in thinking about metaphysical kinds. His contribution, thus, addresses the meta-ontological issue of how epistemology and ontology are – or should be – related to each other. Reydon's chapter ends by pointing to some problems that arise specifically for theories of artefact kinds and do not seem to arise for theories of natural kinds.

In Chap. 9, Beth Preston also asks how we might try to get to a theory of artefact kinds and answers that we should look at "folk" classifications of artefacts. She addresses an issue that is of wider concern than the metaphysics of technical artefacts proper, since it raises the problem of how the philosophy of artefacts is (to be) related to other disciplines studying artefacts, in particular "folk" artefacts, like psychology and anthropology. In her opinion the traditional revisionary or descriptive approaches in the metaphysics of artefacts face a serious problem because they are disconnected from other disciplines studying artefacts, and this gap blocks the metaphysics of artefacts from moving forward. She calls for a new kind of descriptive metaphysics, which as part of an interdisciplinary programme called ethnotechnology may be able to bridge the gap and to connect the metaphysics of artefacts to the work of naturalistic social scientists studying artefacts. One thing that this new metaphysics of artefacts would have to take into account is the fact that ontologies may vary over different communities, resulting in various folk ontologies.

Part III of the book to some extent moves away from issues in theoretical philosophy and turns to ontological issues about technical artefact kinds in engineering practice. Now the focus is on issues regarding the ontologies of engineering and on attempts to develop formal ontologies for technical artefact kinds in the field known as "ontology engineering". In Preston's terminology we are dealing here with issues about particular folk ontologies, the "folk" being particular groups of engineers in their everyday working environment. How do engineers classify technical artefacts and on what grounds? What are the criteria to be imposed on classes of technical artefacts in order for these classes to be candidates for ontologically real kinds?

In Chap. 10, Wybo Houkes and Pieter Vermaas argue that debates in the metaphysics of artefacts typically start from the observation that technical artefacts result from intentional production and then focus immediately on the issue whether this "mind-dependence" leaves artefact kinds admissible as real kinds. They identify two preliminaries to this admissibility problem, namely, which productive activities and whose purposes determine the classification of technical artefacts and which, if any, of these purpose- or activity-based classes may be considered as singling out real artefact kinds. They propose to resolve both preliminary issues by first reconstructing which productive activities are involved in contemporary artefact production and then systematically deriving which classes can be discerned by these activities. The result is a rich "folk metaphysics" of the engineered world, with different classification systems that may correspond to artefact kinds, each with their respective constitutive intentions. After presenting this folk metaphysics, they argue for selecting one of these classification systems, that of "described products", as singling out real artefact kinds. In order to justify this choice, they make explicit their criteria for selecting this system by giving a set of desiderata for real artefact kinds.

In Chap. 11, Nicola Guarino takes the way engineers and technicians speak about technical artefacts and their components as his point of reference. When engineers or technicians speak of technical things, they tend to ascribe a genuine ontological status to their "creatures" even if they do not have a physical presence. This seems to be a systematic phenomenon in the case of missing functional components. Suppose, for instance, that the right headlamp of a car has been smashed in an accident. The lamp is not there anymore, but still a technician would say that a particular cable (perhaps a bit melted) connects to the right headlamp. What does the term "the right headlamp" refer to in this case? The main purpose of this chapter is to offer an ontological account of the different things that engineers and technicians have in mind when they talk about situations like the one above, concerning artefacts and their functional components. Guarino offers an ontological account of artefactual objects and their functional components in which both these categories are distinctly related to intentions: artefactual objects are physical objects that essentially comply with a set of design specifications, and their functional components may be either physical or virtual components that play specific roles in artefactual objects.

The final contribution to this volume, Chap. 12, may be taken, in part, as an empirically informed case study of the ontology of artefacts of a particular folk group, namely, petroleum engineers. Eric Kerr presents a case study supporting the hypothesis that the distinction between natural, social and artificial kind terms has heuristic, rhetorical and ontological value, a fact neglected by those who argue that we should abandon such distinctions. He argues that we should not think of these distinctions as airtight and immune to problematic examples, but as working oppositions: distinctions which have pragmatic value within a particular community. Drawing on the collectivist research tradition, he uses the distinction between natural, social and artificial kind terms to diagnose practices of referencing and classifying tools in petroleum engineering. He concludes that the collectivist account of technical artefacts provides a viable methodology for approaching the ontology of tools and of technical artefacts in general and in the domain of petroleum engineering in particular.

In closing this survey of the content of the book, we would like to point out that the contributions about the epistemological (Reydon) and ethnographical (Preston) approaches to ontology in Part II may offer an interesting way to connect the contributions in Parts I and III. The contributions in these two parts appear quite different in scope, aim and method. This is, among other things, reflected in the fact that the notion of (artefact) kind plays a central role in the analyses of Part I, whereas the notion of (artefact) classification takes centre stage in Part III. This raises the question about the relevance of analyses of how engineers classify technical artefacts for ontological analyses of artefact kinds and vice versa. More in particular, this raises questions about how work on formal ontologies and in ontology engineering, fields that take classifications of items in specific domains as their starting point, is related to traditional philosophical work on the ontology of artefact kinds. Are we dealing here with two distinct conceptual enterprises or are they somehow related to each other?

One way to construct a bridge between these two fields is by assuming that the notion of kind is closely related to the notion of classification. Classification of items (objects, processes, phenomena, events…) may be considered the alpha and omega of the natural sciences, as well as other domains of science and technology; the classifications of plants and animals in natural history may be taken as the beginnings of modern biology and the table of elementary particles in physics as a classification scheme that is based on the most recent physical theories. The ordering of items in classes and subclasses is a way of expressing (part of) our knowledge about the items and the relations between these items. The concepts that are used in a field of investigation correspond with these classes and with the relations between these classes. In formal ontologies and in ontology engineering, attempts are being made to develop formal representations of our knowledge of various domains by developing representations of these concepts and their relations in formal languages. Now, if we approach ontology from an epistemological perspective, as proposed by Reydon, or from an ethnographic perspective, as proposed by Preston, then the main difference between the ontologies so developed and the ontologies developed within formal ontologies and in ontology engineering appears to reside in the use of formal languages in the latter. In both approaches, ontologies are developed on the basis of the best knowledge available; the only difference between the two approaches concerns the use of formal tools. In each of these approaches, one will have to deal with identity problems (with regard to kinds or with regard to classes). And in both approaches one will have to deal with questions about which kinds or classifications are real and what the reality of a classification consists in. Formal ontology and ontology engineering are therefore not only dealing with concepts and relations between concepts, but, as Barry Smith has argued, go "beyond concepts"; the ontologies developed are ways of representing reality (Smith 2004). From this perspective the contributions in parts I and III share a common problem, namely, how technical artefact kinds or classes fit into the ontological structure of the world. If that is indeed the case, the time may be ripe for a fruitful confrontation of work done in both fields; this volume is intended as a first step in that direction.

References

Ayers, M. (1991). *Locke* (Ontology, Vol. 2). London/New York: Routledge.

Baker, L. R. (2007). *The metaphysics of everyday life*. Cambridge etc.: Cambridge University Press.

Dipert, R. R. (1993). *Artifacts, art works, and agency*. Philadelphia: Temple University Press.

Gibbard, A. (1975). Contingent identity. *Journal of Philosophical Logic, 4*, 187–221.

Goodman, N., & Leonard, H. S. (1940). The calculus of individuals and its uses. *The Journal of Symbolic Logic, 5*, 45–55.

Hilpinen, R. (1992). Artifacts and works of art. *Theoria, 58*, 58–82.

Hilpinen, R. (1993). Authors and artifacts. *Proceedings of the Aristotelian Society, 93*, 155–178.

Jubien, M. (1993). *Ontology, modality, and the fallacy of reference*. Cambridge etc.: Cambridge University Press.

Margolis, E., & Laurence, S. (Eds.). (2007). *Creations of the mind: Theories of artifacts and their representation*. Oxford/New York: Oxford University Press.

Merrick, T. (2001). *Objects and persons*. Oxford/New York: Oxford University Press.

Rea, M. C. (Ed.). (1996). *Material constitution: A reader*. Lanham: Rowman & Littlefield.

Smith, B. (2004). Beyond concepts: Ontology as reality representation. In A. C. Varzi & L. Vieu (Eds.), *Formal ontology in information systems* (pp. 73–84). Amsterdam: IOS Press.

Thomasson, A. L. (2007). Artifacts and human concepts. In S. Margolis & E. Laurence (Eds.), pp. 52–73.

Unger, P. (1979). There are no ordinary things. *Synthese, 41*, 117–154.

Van Inwagen, P. (1990). *Material beings*. Ithaca NY: Cornell University Press.

Wiggins, D. (1980). *Sameness and substance*. Oxford: Blackwell.

Part I
Artefact Kinds and Metaphysics

Chapter 2
How Real Are Artefacts and Artefact Kinds?

E.J. Lowe

Abstract It is often difficult to specify, in a principled way, determinate existence and identity conditions for individual artefacts of many familiar kinds, and this fact threatens to undermine realism with regard to artefact kinds and their members. However, it may be argued that at least some kinds of artefacts, especially those that may be categorized as kinds of *machine*, are not vulnerable to these problems. The underlying reason for this is that, in the case of such artefacts, we can identify specific mind-independent laws governing their characteristic modes of activity, which thereby determine their persistence conditions in a nonarbitrary fashion.

Keywords Machines • Natural laws • Ontological categories • Sortal persistence conditions • Utensils

2.1 Introduction

Artefacts and their kinds may be, in a certain sense, *mind-dependent* entities, without this necessarily implying that they are fictional or unreal. Nevertheless, it does not seem sufficient for the reality of an artefact kind and particular instances or members of that kind merely that we should possess a *concept* of that kind whose application conditions are sometimes satisfied. A problematic feature of many such concepts – for example, the concepts *table*, *ship*, *bridge*, *typewriter*, *automobile*, and *computer* – is that they may not seem to be supplied with well-defined or well-grounded *persistence conditions* for the particulars supposedly falling under them. By contrast, where *natural* kinds are concerned, natural *laws* governing those kinds seem to determine the persistence conditions of their instances. A striking illustration of the problem is

E.J. Lowe (✉)
Department of Philosophy, Durham University, 50 Old Elvet, Durham DH1 3HN, UK
e-mail: e.j.lowe@durham.ac.uk

M. Franssen et al. (eds.), *Artefact Kinds: Ontology and the Human-Made World*, Synthese Library 365, DOI 10.1007/978-3-319-00801-1_2,

provided by a recent Turner Prize exhibit, entitled 'Shedboatshed' by the artist. In this case, the artist reconfigured a wooden shed as a functioning wooden boat, sailed it along a river, and then reconfigured it as a shed once more. This immediately prompts the following question. Was this just *one* artefact which successively underwent two radical metamorphoses, or a sequence of *three* numerically distinct artefacts – a shed followed by a boat followed by another shed? And if it was just one artefact, then what *kind* of artefact was it? A *shedboatshed*, as the artist suggested? But that could hardly be described as a familiar artefact kind: it was indeed remarkably ingenious of the artist to come up with the idea. Or was this instead a case of one artefact, *a shed*, temporarily ceasing to exist while its parts were appropriated by another, *a boat*, and subsequently coming back into existence again? We seem to be faced with a difficult choice between saying that one, two, or three numerically distinct artefacts were involved in this situation. All three answers can't be correct, since they contradict each other. Less extreme but equally troubling examples are abundant in the domain of artefacts, calling into question the reality of that domain. For, if the cardinality of a domain is not determinate, the reality of the entities supposedly belonging to it is put in doubt. To quote the famous words of Quine (1969): 'No entity without identity'. In this chapter, I shall attempt to steer a middle path between a complete rejection of the reality of artefacts and their kinds, on the one hand, and an unjustifiably luxuriant ontology of artefact kinds and their members on the other.

2.2 Kinds, Categories, and Sortal Persistence Conditions

Artefacts – that is to say, *particular* artefacts – are, if they are anything at all, *individual concrete objects*: property bearers which exist in space and persist through time. (I am setting aside here putative examples of *abstract* artefacts, such as musical scores, conceived as *types* rather than tokens.) As such, however, they are and must be instances or members of certain *kinds*. This is because, I think, no clear sense can be made of the notion of a perfectly 'bare' particular (see further Lowe 2009: pp. 14–16). A bare particular, if there could be such a thing, would be a concrete individual – a particular entity existing in space and time – which was capable of 'supporting' and in that sense *having* properties, but with no restriction whatever on the range of properties that it could, at different times, support and hence with no restriction whatever on its *persistence conditions*. On this view, the 'same' bare particular might support, at one time, all the properties that one would normally associate with, say, a *tiger*, but at another time all the properties that one would normally associate with an *electron*. However, I would urge that, with the notion of a particular so absolutely *Protean*, whose identity over time is consequently so completely unconstrained, we lose all grip on the idea that it is genuinely something with an *identity* at all. And, without an *identity*, it simply isn't clear that we can seriously suppose that we really have an *entity*. Once again, as Quine so succinctly put it, 'No entity without identity'.

Now, concrete particulars of different kinds evidently possess, very often, different *existence and identity conditions*. The existence and identity conditions of *mountains*, for instance, clearly differ from those of *monkeys*. This is because mountains are a kind of *geological formation*, whereas monkeys are a kind of *living organism*. The terms 'geological formation' and 'living organism' denote different *ontological categories* – different *sub*categories of the higher-level category *individual concrete object* (or, to use an older terminology, *individual substance*). Here, by implication, I am distinguishing between *categories* and *kinds*. How exactly to draw this distinction is a matter of some controversy (see, for example, Jan Westerhoff 2005), but that it needs to be drawn seems clear. It also seems clear that, by no reasonable standard, could we say that terms like 'mountain' and 'monkey' denote *ontological categories*: they are simply far too specific for that. Ontological categorization operates at a far higher level of abstraction than do the taxonomic practices of special sciences, such as geology and biology. In due course, I shall have more to say about the category/kind distinction, but it is no part of my present purpose to lay down a hard-and-fast criterion for the distinction (see, however, Lowe 1998, pp. 178–185 and Lowe 2006a, pp. 3–8 for detailed proposals).

What I shall call the *sortal persistence conditions* of an individual object clearly *depend upon its kind*. These are the conditions that are necessary and sufficient for its persistence *as an instance of the kind*. However, it is a matter for debate whether or not *sortal essentialism* is true (a doctrine espoused, for instance, by David Wiggins 2001). A sortal essentialist claims that an individual's sort or kind is *part of its essence*, with the consequence that its sortal persistence conditions coincide with its individual *identity* conditions. (In this connection, it is important not to confuse an individual's *sort* or *kind* with one or more *phases* that it might exhibit in the course of its natural development: thus, a *tadpole* is just a frog in its larval phase, not a different *kind* of living creature.) A non-sortal essentialist maintains, by contrast, that individuals can – in the sense of 'can' in which it expresses *metaphysical* rather than merely natural or physical possibility – survive a change even of their 'highest' sort or kind, but *not* of their ontological category. (Clearly, sortal or kind terms of varying degrees of generality can be applied to one and the same individual: thus, a frog also belongs to the 'higher' kind *amphibian*.) However, interesting though it is, I shall set aside this dispute as far as possible for present purposes, although my own preference is for non-sortal essentialism (see Lowe 1998, pp. 55–56, 185–187). In fact, I shall for the most part assume sortal essentialism as a working hypothesis, because many philosophers have an explicit allegiance to it and even more would probably regard the implications of its denial as implausible. Consequently, I shall for the most part assume that an object's sortal persistence conditions coincide with its individual identity conditions.

2.3 How Are Sortal Persistence Conditions Grounded?

The term 'artefact' certainly seems to denote an ontological category – a *medium-level* one, much like 'geological formation' and 'living organism'. (As we shall see shortly, there are plausibly also *sub*categories of artefact, such as *utensil* and *machine*.) Let us tentatively assume that this is indeed so. Then what can we say about *artefact kinds* and the associated sortal persistence conditions governing the individual objects that belong to them? *Prima facie*, there *are* such kinds, such as the kinds *table*, *ship*, *bridge*, *typewriter*, *automobile*, and *computer*: certainly, all of these terms appear to express kind *concepts*. Some of these putative kinds seem, moreover, to be *sub*-kinds of higher kinds, just as *frog* is a sub-kind of *amphibian* and *monkey* is a sub-kind of *mammal*. For instance, it seems natural to say that ships and automobiles are both sub-kinds of *transport*. Now, it might be thought that, by the same token, automobiles and computers, say, are both sub-kinds of *machine*. I agree that this sounds perfectly correct, as far as it goes, but I would nonetheless resist saying that 'machine' denotes a high-level artefact *kind*. I think it is more plausible to say that it denotes a sub*category* of artefact. Here again we come up against the problem of saying exactly *how* and *where* to draw the line between categories and kinds. However, I shall once more, for present purposes, rely more on intuition than argument to determine my answers to such questions, since it is more important that the distinction be recognized than that an agreed way of drawing it in all cases be settled upon.

Now, what *grounds* the distinctive existence and identity conditions governing the members of different kinds – where, by such conditions, I mean those necessary and sufficient for the *existence* of any given member of the kind and for the *identity or distinctness* of any two such members? Here we are immediately confronted with the fact that *natural* kinds – such as the kinds *monkey* and *mountain* – are presumably wholly *mind-independent*, whereas artefact kinds are plausibly not. The existence and identity conditions of the members of artefact kinds plausibly carry reference to human or other intelligent *artificers* and *utilizers* of them: the people who make and use artefacts for certain purposes, or to serve certain functions. Without the *intentions* of such people, it seems, there could *be* no artefacts. However, this fact doesn't render artefacts mind-dependent in the way, for example, that *fictional characters* plainly are: artefacts don't exist only because they are *believed* or *imagined* to exist. (Here it might be objected that *money* is an artefact and yet plausibly does exist only because it is believed to exist. To this, however, I would reply that we need to distinguish between money in the sense of *currency* – the pound sterling or the US dollar, for example – and money in the sense of *specie*, in the form of coins and the like. The latter are artefacts, but not the former. Money in the sense of currency is instead an institutional or social construct and indeed exists only so long as people have faith in it.) The key point, it seems, is that artefacts are only *causally* mind-dependent, not *constitutively* mind-dependent. An artefact could survive the demise of all intelligent life, but no artefact could *come into existence* without intelligent life existing, since only intelligent beings can create artefacts.

Before proceeding further, I want to question what I shall call *conceptualist realism* as a general account of sortal existence and identity conditions (for a fuller critique, see Lowe 2008). This is the view that *K*s exist – where '*K*' is any *sortal* or *kind* term – just so long as we or other intelligent beings have a coherent *concept* of *K*s, whose application conditions turn out to be *satisfied by the world*. (Who holds this view? It seems that David Wiggins (2001) does and perhaps also Amie Thomasson

(2007), although I am more concerned to question the view itself than to challenge any particular adherent of it. For a critique of Wiggins's version of the view, see Lowe 2006b.) I don't think that this view can be correct as a *general* account, on pain of global antirealism. This is because it seems that we can speak non-vacuously of a sortal concept's being 'satisfied by the world' only if some *things* in the world have properties and stand in relations demanded by the application conditions of the sortal concept in question. (I shall consider the consequences of denying this in a moment.) But that presupposes that *the things in question* have existence and identity conditions too, which need to be explained and grounded in turn.

For instance, a conceptualist realist might say that the application conditions of the concept of a table are such that for a *table* to exist is just for some pieces of wood or metal to be *thus-and-so configured* – broadly speaking, in such a way as to provide a stable means of horizontal support for medium-sized material objects – and that it is similarly a conceptual truth that the table *persists* just so long as this configuration or arrangement of its material parts continues to obtain (allowing, no doubt, for some gradual replacement of a few individual parts over time, should the table occasionally need to be repaired). But what, now, about the existence and identity conditions of those material parts themselves – the pieces of wood or metal? If we try to tell exactly the same story about them too, then we shall obviously have set out upon a regress which either has no terminus or else is terminated by entities whose existence and identity conditions cannot be accounted for in the same way, in terms of certain of our concepts being 'satisfied by the world'. If the latter, then the account is not, after all, perfectly general. If the former, then no *account* has really been given, because the crucial question has been permanently deferred. On the other hand, should the conceptualist realist try instead to appeal to some notion of a sortal concept's being 'satisfied by the world' which does not require articulation in terms of some *specific entities* in the world possessing certain properties and standing in certain relations, then, I suggest, we are left with a form of global antirealism. For then we are operating with a notion of a 'world' that is not already pre-conceptually articulated into objects bearing properties and standing in relations to one another, but is just an 'amorphous lump', in the vivid phrase of Michael Dummett (1981, p. 563).

Some metaphysicians invoke, as a species of *filter* on overindulgent ontologies of sorts or kinds, some version of the 'Eleatic principle' or 'Alexander's dictum', according to which kinds are real only if their instances or members have *independent causal powers*. Trenton Merricks (2001), for instance, dismisses the reality of *baseballs* and other such composite nonconscious objects for this reason, maintaining that they have no such powers. According to him, everything that a baseball supposedly *does*, causally – such as break a window – is really done by certain *material particles arranged baseball-wise 'acting in concert'*. This, however, is surely too austere a doctrine, if indeed it really works at all as a way of reducing our ontological commitments. (I make this qualification because I have serious doubts as to whether talk of 'particles arranged baseball-wise acting in concert' can really be understood other than merely as a circumlocution for talking about *baseballs acting*: see further Lowe 2003.) Even so, there is something appealingly restrained about this approach to questions of ontology, at least compared with the apparent luxuriance of conceptualist realism. It is surely true to say that a kind and its members have to *earn their keep* to deserve a place in our ontology. But I think that, in the case of *natural* kinds, at least, we can pinpoint what it is that enables them to do this, without resorting to the austerity of the position described earlier.

In the case of natural kinds, we can appeal to *natural laws* as providing the grounds of the existence and identity conditions of their respective members. For instance, there are laws governing the persistence of *monkeys*, telling us how they are generated, develop, and ultimately die – laws that are significantly different from those governing, say, *mackerel* (a kind of fish) or *mosquitos* (a kind of insect). Such developmental laws are not, of course, comparable to the exceptionless laws of fundamental physics: they are, in the current jargon, *ceteris paribus* laws – but they are, nevertheless, empirically well-confirmed generalizations concerning particular biological species, correctly informing us, for instance, how tadpoles normally develop into mature frogs. That such laws are very probably susceptible to explanation in terms of more fundamental laws, say of genetics and ultimately of physics, is no threat to their status as laws. Even *geological formations* are governed by natural laws specific to their kind: there is, after all, a science of *mountains* – orology – telling us how they are generated, develop, and are ultimately eroded into nonexistence. No mountain can change naturally into a geological formation of a quite different kind, such as a *crater* or a *canyon*: and this is because the persistence conditions of mountains, as determined by the natural laws governing them, are quite different from those of craters or canyons. It is true that a mountain can be naturally transformed into an *island*, just by becoming surrounded by water: but that, very arguably, is just a reason for saying that an island simply *is* a mountain that is surrounded by water, rather than a geological formation of a quite different kind.

2.4 Problems with the Identity of Artefacts

Now, with *artefact* kinds, such matters of persistence and identity undoubtedly appear to be much more problematic, as we have already remarked. Can a *table* 'morph' into a *chair*? It seems so – in fact, it can do so very easily. Indeed, whether a table has done so may on occasion apparently lie simply 'in the eye of the beholder' or, at least, of the *user*. A table can certainly be *used as* a chair – or as a *stool*, if one prefers, on the grounds that a chair must have a back. Does that *make* it a chair? If so, has it then ceased to be a *table*? Moreover, it seems that an artefact of one kind can, with a little ingenuity, often be *reconfigured* so as to turn it into one of another quite different kind – for instance, a chair might in this way be transformed successively into a stool, a birdhouse, and then a table (Burke 1980). The recent Turner Prize exhibit mentioned earlier, 'Shedboatshed', is but an extreme – if also artistically intriguing – example of this seemingly almost unlimited plasticity of artefacts and their kinds. And although in science fiction one might imagine something similar in the domain of animals and plants, in science *fact* the possibilities appear to be much more constrained, even with current advances in genetic engineering.

Why exactly, though, might these considerations be thought to present a threat to the *reality* of artefacts and artefact kinds? As I indicated earlier, the answer is really very simple. It is because existence and identity conditions determine the answers to questions concerning the *cardinality* of a domain of entities – and if such questions *don't* have determinate answers with respect to a certain domain, then the reality of that domain is immediately open to dispute. To quote Quine once more: 'No entity without identity'. If there is no *fact of the matter* as to whether, say, a certain table has *ceased* to exist and been *replaced* by a chair, or has *become* a chair, perhaps while *remaining* a table, then it simply isn't clear whether we are dealing here with just *one* thing or with *two*. After all, we may think that a table can survive being *folded up*, despite the fact that in this condition it cannot actually *function* as a table, so why shouldn't it be able to survive being 'folded up' in the shape of a *chair* and actually be *used* as a chair in that condition – or indeed 'folded up' in a host of other ways, so as to serve alternatively as a *dog kennel*, a *birdhouse*, or pretty much *any* other kind of artefact? (Of course, philosophers have been generating puzzles about artefact identity ever since the ancient problem of the Ship of Theseus was first conceived. But what is different about the present sort of puzzle is that it involves more than one *artefact kind*, whereas the problem of the Ship of Theseus involves only one such kind – the kind *ship*. I discuss the latter problem in some detail in Lowe (1983).)

Are there any principled constraints *at all* that can be applied to restrict such possibilities, comparable to those that operate in the domain of natural kinds? It is perhaps worth noting here that abandoning *sortal essentialism* for artefacts does not really promise to help us very much in resolving this difficulty. In the case of *living organisms*, for instance, a denial of sortal essentialism, with the consequent admission of the metaphysical possibility of an individual organism's surviving a radical metamorphosis from, say, being a *frog* to being a *dog*, still leaves us with a firm grip on the *identity* conditions of individual organisms, which in all cases seem to turn on the equivalence relation of *sameness of individual life* – the relation that John Locke (1975 [1690]: II, XXVII, 4) famously regarded as providing a criterion for the identity over time of living things (for discussion, see Lowe 2009, pp. 107–110). But it is far from clear that there is any single comparable equivalence relation that can be appealed to in the case of all individual *artefacts*, to underpin *their* identity conditions in a uniform way, without regard to any artefactual sort or kind to which they might be deemed to belong.

2.5 The Contrast Between Machines and Utensils

Do such doubts, then, call into question the reality of *all* putative kinds of artefacts and their members? I strongly suspect not. Some artefact kinds are pretty much *defined*, it seems, by reference to the *use* that we or other intelligent beings put them to, that is, by reference to the *functions* that we regard them as serving, or as being apt to serve, on our behalf (for, of course, they can be *apt* to serve certain functions while not *actually* serving them, as in the case of a folded-up table or a packed-away tent). Into this category of artefacts – which I propose to call *utensils* – fall tables, chairs, tents, cooking pots, knives, and hammers, to name but a few. I seriously doubt whether *these* supposed things are genuinely 'real'. I see no compelling reason to say that, when a piece of wood and a piece of metal are put together in the shape of a knife or a hammer, an *object* of a new kind comes into existence. Rather, some pre-existing objects are just brought into a certain relationship that renders them useful to serve some humanly desired function, such as cutting or flattening. However, there is at least one subcategory of artefacts that seems different and special, with respect to the claims of its members to be regarded as real. This is the category of *machines*, one paradigm example being that of an *engine* or *motor*. Another equally good example is that of a mechanical *clock*. Why are these different and special in this respect? My answer is that the members of such machine kinds all seem to embody a distinctive *unifying principle of activity* which is constrained by *sortal-specific laws*. The laws in question are, moreover, distinctively laws of *engineering*, not mere laws of 'natural science' or 'physics', in the entirely indiscriminate sense in which the latter concern the behaviour of physical systems quite generally, at all scales of size and mass from that of fundamental particles upwards. (A simple example of such an engineering law would be the principle explaining the action of a *centrifugal governor* in a steam engine. An even simpler and more familiar example would be the law of the *pendulum*. In each case, the law in question concerns a certain type of potential *machine component* and helps to explain how it can perform its characteristic role in any machine in which it is incorporated as a working part.) But in another and broader sense, of course, these sortal-specific engineering laws are still undoubtedly *natural* laws, where such laws are to be contrasted, for instance, with mere human customs or conventions, and have an objective foundation in mind-independent reality.

A Martian, visiting Earth after all human life had been extinguished, could surely recognize in a working piston engine an object that *does something*, by its own very nature – for example, *the pistons turn the crankshaft*, *the valves regulate the pressure in the pistons*, and so on. The Martian would not need to know of any *use* to which humans had put such engines in order to recognize this fact. Similar observations apply to something like a mechanical pendulum clock: *the pendulum regulates the turn of the cogwheels*, *the weights keep the pendulum swinging*, and so forth. By contrast, a table or a hammer, it seems, *does nothing at all* by its own very nature: it is only configured in a way that makes it apt for *us* to use, so that *we* can do something *with it*. In this case, *we alone are the doers*, not the table or hammer. Of course, we *can* also 'use' an engine in this sense – for instance, we can use it to help us to transport heavy goods, by employing it to drive a wheeled vehicle. But that is only because of what the engine *itself* is capable of doing, by its own nature – namely, *driving the wheels of the vehicle*. The key point is that the engine *has* a 'nature', which is governed by natural law. Tables and hammers are, of course, *subject to* natural laws: if a table is dropped from a height, it will fall, being subject to the law of gravity. But tables and hammers are not, it seems, *subjects of* natural law: there are no laws governing what *they* do – for they do *nothing* of themselves, by their own nature, and in that sense *have* no 'nature'. To explain a table's fall, no reference to the fact that it is a *table* is needed, since the fall is entirely explained in terms of the nature of the material parts of the table (their weight, shape, air resistance, and so forth). Of course, it might be said that the same is ultimately true of machines too: that all of their activity is ultimately determined by laws governing the material particles that compose them. That may or may not be so, but in any case this ignores the distinction between *special-science* laws and *fundamental physical laws*. There is no special science of tables or hammers, comparable to the science of engineering that applies to machines, and no corresponding special-science laws, comparable to the laws of engineering.

Should we say, then, that machines constitute a *natural kind*, or at least divide into various different natural kinds, such as *piston engines* and *pendulum clocks*? No, I think not, because they are not *generated* – brought into being – by mind-independent nature, only by intelligent beings, who design and manufacture them. Nonetheless, they do have a real 'nature' of their own, because their creators endow them with a unifying principle of activity that is governed by mind-independent laws of action, specific to their sort or kind: *engineering laws*. Such a 'principle of activity' is analogous to the sort of *life principle* that unifies the activity of any living organism, coordinating the activities of all its parts in order to preserve the structure and activity of the organism as a whole. One mark of a being that genuinely has such a real, unified 'nature' is the fact that we can meaningfully talk about the *malfunctioning* of one of its parts: this happens when a part ceases to work in a way that is conducive to the characteristic activity of the whole. Thus, just as organisms can suffer from *diseases*, so machines can suffer from *breakdowns*. A knife or hammer can of course be *broken*, but this is not like the 'breakdown' of a machine, which involves a critical impediment to its characteristic way of working.

2.6 Concluding Remarks

Here I reach a tentative conclusion concerning the reality of artefacts and artefact kinds. I believe that we should adopt a *moderate* rather than an *exuberant* realism regarding artefact kinds and their members. The term 'artefact' does not, on closer examination, really seem to denote a genuine *ontological* category, because at least some of our talk about artefacts should probably be interpreted antirealistically. However, *not all artefacts are equal* in this respect. There *are* artefacts and artefact kinds that are very arguably fully real – certainly, at least as real as *natural* kinds and their members are. Genuine machines, embodying a distinctive unifying principle of activity, seem to have persistence conditions that are lawfully determined, quite as much as living organisms do. When such a principle of machine activity ceases to be realized in any given case, the machine in question is *destroyed* and thereby ceases to exist. Of course, a machine may continue to exist even while it is not actually working, but in like manner a living organism may be cryogenically preserved and subsequently be 'brought back to life'. This analogy between the working of a machine and the life of an organism seems sufficiently close to place machines and organisms on the same side of the realist/antirealist divide, so that no philosopher who is a realist about organisms should be an antirealist about machines. In both cases, it appears that we have a full enough account, in mind-independent terms, of the grounds of their existence and identity conditions to counter the concerns about cardinality that feed the antirealist's doubts. But not so in the case of *utensils*, it seems.

Of course, this then raises the question of how sharp the machine/utensil distinction can credibly be said to be. To which I reply that, while there may indeed be borderline cases, so too are there between living and nonliving things, and that we should not allow the existence of borderline cases to undermine our belief in real distinctions. After all, red is really distinct from orange and yellow, despite the fact that some shades of orange are borderline cases of red, while others are borderline cases of yellow. Finally, I should stress that I am not urging that machines are the *only* real artefacts. I am content to allow, for instance, that *works of art* may well qualify as real artefacts too. Indeed, it seems plausible to say that what so easily persuades some people to regard the exhibit Shedboatshed as being a *single* object persisting through a series of radical transformations is that, in doing so, they are conceiving of it as being a unique work of art, rather than as an artefact that has any essential practical utility or function, or any characteristic manner of *working*.

References

Burke, M. B. (1980). Cohabitation, stuff and intermittent existence. *Mind, 89*, 391–405.

Dummett, M. A. E. (1981). *Frege: Philosophy of language* (2nd ed.). London: Duckworth.

Locke, J. (1975 [1690]). In: P. H. Nidditch (Ed.), *Essay concerning human understanding*. Oxford: Clarendon Press.

Lowe, E. J. (1983). On the identity of artifacts. *The Journal of Philosophy, 80*, 220–232.

Lowe, E. J. (1998). *The possibility of metaphysics: Substance, identity, and time*. Oxford: Clarendon.

Lowe, E. J. (2003). In defense of medium-sized specimens of dry goods. *Philosophy and Phenomenological Research, 68*, 704–710.

Lowe, E. J. (2006a). *The four-category ontology: A metaphysical foundation for natural science*. Oxford: Clarendon.

Lowe, E. J. (2006b). Is conceptualist realism as stable position? *Philosophy and Phenomenological Research, 71*, 456–461.

Lowe, E. J. (2008). How are identity conditions grounded? In C. Kanzian (Ed.), *Persistence* (pp. 73–89). Frankfurt: Ontos Verlag.

Lowe, E. J. (2009). *More kinds of being: A further study of individuation, identity, and the logic of sortal terms*. Malden/Oxford: Wiley-Blackwell.

Merricks, T. (2001). *Objects and persons*. Oxford: Clarendon.

Quine, W. V. (1969). Speaking of objects. In W. V. Quine (Ed.), *Ontological relativity and other essays* (pp. 1–25). New York: Columbia University Press.

Thomasson, A. L. (2007). *Ordinary objects*. Oxford: Oxford University Press.

Westerhoff, J. (2005). *Ontological categories*. Oxford: Clarendon.

Wiggins, D. (2001). *Sameness and substance renewed*. Cambridge: Cambridge University Press.

Chapter 3
Artifacts and Mind-Independence

Crawford L. Elder

Abstract Realists maintain that the only material objects there are those that exist mind-independently. In this chapter I argue that realists can consistently maintain that artifacts of many, many familiar kinds really exist. I end by suggesting that realists can even affirm the reality of artifacts that are not material objects at all – "cultural artifacts" such as gender roles or codes of etiquette. Exactly what are the reasons for *doubting* whether realists can consistently affirm the reality of artifacts? One reason that probably motivates some doubts is that artifacts are things that we create. A more substantial reason, as we shall see, is that artifacts appear to belong to diverse kinds, each essentially characterized by a distinctive function or functions. The function that characterizes the artifacts in a given kind appears to be fixed by how we act and think in regard to those objects; so if that function is essential to those artifacts, we are responsible for those objects' having at least one of their essential properties. By itself, I shall argue, even this reason for suspecting that realists cannot afford to affirm artifacts is less substantial than at first it appears. But it may appear inseparable from a third and truly substantial reason for doubting whether realists can affirm artifacts. This reason is that, since we fix the functions of artifacts by our actions and thoughts, we are in a privileged epistemic position with respect to (at least some of) their essential properties – a position too privileged to leave room for the idea that those artifacts exist mind-independently. This worry is acute in the case of "capitalist artifacts," by which I shall mean artifacts produced in the relations of production that Marx called "the commodity form of production." It is even more acute in the case of "cultural artifacts." But this worry can nevertheless be dispelled, I shall argue.

Keywords Artifacts • Commodities • Epistemic privilege • Mind-independence • Realism

C.L. Elder (✉)
Philosophy Department, Unit 1054, University of Connecticut,
344 Mansfield Road, Storrs, CT 06269-1054, USA
e-mail: crawford.elder@uconn.edu

M. Franssen et al. (eds.), *Artefact Kinds: Ontology and the Human-Made World*,
Synthese Library 365, DOI 10.1007/978-3-319-00801-1_3,

3.1 Functions as Generated by the Activities of Conscious Beings

If there really are artifacts in the world, they are objects which conscious subjects make, by virtue of their conscious activities. Does it follow that artifacts would be mind-dependent objects – in the very sense in which realism denies that there are any mind-dependent objects? In the absence of extra premises, that conclusion does not appear to follow. By virtue of our industrial activities and our agricultural practices, we create large amounts of "greenhouse gases." But it seems not to follow that realists cannot afford to admit that these large volumes of greenhouse gases really exist.

We invented rayon and manufacture it to this day. We constructed the Verrazano-Narrows Bridge. Does it follow that realists cannot afford to admit that there exists such a textile as rayon or such an object as the Verrazano-Narrows Bridge? Exactly what realism *does* deny is a topic to which I shall later return. But offhand, it seems as if there is no reason why realists cannot admit that there are materials and objects that causally depend upon our physical activities. Certainly realists who are also naturalists *do* claim that there are such materials and such objects; a large part of why natural selection brought us about, these naturalist realists would say, is precisely our predecessors' repeated historical successes at fashioning materials and making objects. The combination of realism with this sort of story about natural selection strikes no one as astonishing or appalling.

On the other hand, the particular examples of rayon and the Verrazano-Narrows Bridge suggest a premise that might, indeed, appear to pose a threat to realism. We invented rayon and manufacture it to this day, with a particular purpose in mind: rayon is supposed to function in certain ways as a fabric. The function of any bridge is to conduct traffic (pedestrian or vehicular) across an expanse of water or unfriendly terrain, and the function of the Verrazano-Narrows Bridge in particular is to conduct vehicular traffic between Brooklyn and Staten Island. Artifacts of any kind are characterized by a particular function or functions. And it is our ways of acting and thinking that appear to underlie these functions. It seems that the producers of artifacts instill in them their functions; or that the users do, by the ways they employ artifacts; or some combination of the two.

Here is one way of focusing this extra premise. Suppose, counterfactually, that conscious subjects were removed from the world's history. Would it follow that there could not exist in the world bridges or textiles – or televisions or shoes? On plausible assumptions about the laws of nature operative in our counterfactual world, and about initial and boundary conditions obtaining in it, there *would* not exist bridges or shoes, textiles or televisions. No nonconscious agents or processes would have been adequate to cause such objects. But set the question about causation aside. Are such things as I have listed tied not just causally to conscious agents, but in some stronger way, is their existence tied by conceptual necessity, or metaphysical necessity, to the existence in the world of conscious agents?

It seems that the answer is Yes. There could have been a massive aggregation of steel looking just like the Verrazano-Narrows Bridge, we apparently should say – since we are setting aside misgivings as to how such an aggregation could have been caused to exist – and yet it still would not have been a *bridge*. Shoes, to be shoes, seem to be tied by something stronger than just causal necessity to the existence of conscious subjects who have feet. In short, it seems that artifacts of any kind are *essentially* characterized by a certain function or functions and can have the functions that they do only in virtue of the ways conscious subjects act and think with regard to them.

Here, then, is a serious reason for doubting whether realists can afford to affirm artifacts: artifacts appear to be essentially characterized by properties – viz. functions – that are mind-dependent.

3.2 Essential Status as Underlain by the Cognitive Practices of Conscious Beings

But there is a higher-order variant of the worry that we are responsible for essential properties, and it threatens the realist's ability to attribute mind-independent reality not just to artifacts, but to any material objects whatever. Realists must address the higher-order worry first. Only afterward can they seek to argue that the functions by which artifacts of each kind are essentially characterized are not, after all, mind-dependent – at least, not mind-dependent in any way that is incompatible with realism.

The higher-order worry arises because of the importance, for realism, of persistence. Any version of realism worth defending must attribute mind-independent existence to objects that neither have always existed nor will always exist. These are objects that begin to exist at a certain time, that continue to exist across certain changes – even if only such trivial changes as change in location or change in age – and that pass out of existence at a later time. In order to attribute mind-independent existence to such perishable objects, I contend, one must attribute to them mind-independent *courses* of existence, careers, that mind-independently begin and end where they do and that mind-independently span such changes as the object undergoes. Now one could indeed imagine a version of realism that made no such claim. This version would say that each perishable object exists mind-independently but that the beginning of its existence, the continuing of its existence across changes, and the ending of its existence are all phenomena that obtain only in virtue of our ways of thinking or talking. But such a position would seem miserably unstable. Each perishable object would enjoy mind-independent existence, but there would be no mind-independent fact of the matter as to *when* it exists, no mind-independent *span* that its existence takes up. As it exists mind-independently, any such perishable object would be without temporal location – and hence would be quite disturbingly incomplete. Meinong believed that there are just such "incomplete objects" – persons who were not born on any particular day, for example, or mountains that do not have any particular height or weight, or ivory spheres that have no particular diameter (Findlay 1963, ch. VI). But it is nearly impossible not to regard such incompleteness as the mark, precisely, of a fictional object. Just so with perishable objects that, supposedly, exist mind-independently, but do not mind-independently exist some*when*.

But the need to attribute mind-independent courses of persistence to objects requires, on plausible assumptions, a mind-independent difference between what it is for properties to be essential to a given object and what it is for properties to be merely accidental. For that is the most natural account of the difference between those events that constitute mere alterations in an object's career and those that amount either to that object's ceasing to be or to its coming into existence. An object ceases to exist, on this natural account, just when it loses properties that are essential to it; it came into existence at just some occasion on which such essential properties came jointly to be instantiated; and the alterations across which it persists are just those in which the object loses one accidental property and replaces it with another (Elder 2011, ch. 3).

This reflection is what creates the higher-order variant of the challenge to realism that we considered in the previous section. There, we considered just the worry that we are the authors of the function characteristic of artifacts of any given kind – a property which, we were supposing, is essential to artifacts of that kind. But what if, even in the case of those properties of objects of which we are *not* the authors – those properties that attach to objects independently of how we think and act with regard to those objects – we nevertheless are the authors of their status *as* essential? Then it is we who are responsible for the fact that loss of certain properties, by any object, amounts to mere alteration, while loss of certain others amounts to a failure of persistence. We are the authors of persistence conditions, for objects. And then courses of persistence are sequences of events for which *we* fix the end points. This is *anti*-realism about persistence.

This, then, is the broader question which realists must address first, before saying anything about the functions characteristic of each kind of artifact: what is it for properties to be essential to the objects in this or that kind, and can their being essential be said to be a mind-independent phenomenon? The question is challenging just because the answer most widely endorsed, in contemporary metaphysics, pictures essential status as determined by us and our conventions. More carefully, what "modal conventionalism" says is this. Associated with any of our sortals (or matter-names) are certain "conventions of individuation" (Sidelle 1989, pp. 67–68, 85; Thomasson 2007a, pp. 38–44, 48–53). Some of these shape the correct answers as to how kinds are individuated, within that family of kinds to which the kind picked out by the sortal belongs – the answers, that is, as to what kinds of sameness unite each such kind across its extent and separate it from other kindred kinds. Other conventions shape the correct answers as to what is involved, not in being a member of this kind or that kind, but in being *the* very member of this kind or that as was present on some earlier occasion (Sidelle 1989, pp. 52–57; Thomasson 2007a, pp. 56–59, 157). Between them the two sorts of conventions fix which sorts of properties it is that an object, or a sample of some natural kind of matter, must carry with it across the span of its existence: they fix persistence conditions. The conventions may indeed incorporate a "blank check" for empirical research to fill in.

For example, our conventions for individuating chemical kinds say that each is united across its extent, and set off from other chemical kinds, by whatever deep-lying microstructural feature it is that empirical science discovers. In the case of the kind *gold*, empirical research fills in the blank check with the entry "the element bearing 79 neutrons in the nuclei of its component atoms." Modal conventionalists can therefore say that we *learn empirically* that any sample of gold must, to go on existing, retain atomic number 79. Even so, the blank check is valid only because *we* signed it. That is, sameness in atomic number makes for sameness in kind, and retention of atomic number amounts to a persistence condition, just because and just to the extent that our conventions *say* it is the *kind* of sameness that plays this role.

But there is also a realist position on what it is for properties to be essential to the members of this kind or that. This position, which I have put forward over a number of years, says that for certain properties to be the ones essential to a given natural kind is for them to be joined by causal connections that make their clustering together, in instance after instance, no accident (Elder 1992, 2004, ch. 2). The causal connections moreover ensure that if, in a different but kindred natural kind, a property occurs which contrasts with one in the original cluster, that contrasting property will entrain the presence of other properties that contrast correspondingly with others in the original cluster. Essential properties are causally "geared together": if the setting on one of the knobs is twisted, the world will display the other knobs as having moved to different settings as well. Importantly, *that* this causal gearing is what essential status amounts to has nothing to do with any conventions we subscribe to, nor to any "referential intentions" that we harbor, nor to anything that is in the heads of contemporary speakers and thinkers. Instead the centrality of causal gearing to essential status derives from what has (on plausible assumptions) underlain, as a matter of objective historical fact, the "survival value" of our sortals – and most crucially of the judgments of kind-sameness and of numerical persistence which our sortals enable us to make (Elder 2011, ch. 1).

So there is an available realist position on what, *in general*, it is for properties to be essential. Now back to that interesting essential feature of the *artifacts* in any given kind, namely function. Is there an account of function such that the function characteristic of artifacts in a given kind figures in this sort of world-given causal gearing? And if there are such realism-amenable functions, do they succumb to the objection that they are mind-dependent? In the next section I answer Yes to the first question; in the subsequent section, No to the second question.

3.3 Etiological Functions and the Natures of Artifacts

There is a great deal of discussion, both within the philosophy of artifacts and within the philosophy of biology, as to what it might be for this or that to be "the function" (or "a function") of a particular artifact or device or trait. There is one account of what a "function" is that does indeed depict the "function" characteristic of some kind – some kind of artifact *or* some kind of biological device – as causally geared together, in the way discussed above, with other properties that are essential to that kind. This is the etiological account of function put forth most notably by Ruth Millikan. I believe, though I will not defend the claim here, that such etiological functions – Millikan's term for them is "proper functions" – are the only sorts of functions that can be said to be geared together with other kind-characterizing properties in the way the realist requires for essential status. What I *will* argue is that *at least* Millikanian proper functions answer to the realist's requirements on essential status.

In order for the members of some kind to be characterized by a proper function, there must be some mechanism or system which copies new members of the kind from ancestor members. This copying mechanism must in principle be capable of varied outputs: it must be the case that if earlier members had differed in some ways from their actual morphology, the mechanism would have produced correspondingly different copies (Millikan 1984, ch. 1). The actual copied members will have a proper function just in case what causally explains their repeated production is some effect that the actual ancestor members brought about. The ancestor members need not have brought this effect about always and may not even have brought it about often. So long as their having brought it about as often as they did is what causally explains continuation of the copying process, then the copied members will have producing that effect as their proper function. (For completeness, I should add that the "causal explanation" here is contrastive: what has to be the case is that the actual morphology gets reproduced, *rather than* some *alternative* historical morphology, as a causal consequence of previous members' having produced, in virtue of having that actual morphology, that characteristic effect – Millikan 1993, p. 38). Intuitively, one can say that producing this characteristic effect is what current members of the kind are "supposed to do": it is "why they are there."

Thus the members of such a kind are, in a broad sense, fashioned by a process of selection. The performance which new members of the kind have been selected for – their "proper function" – is one which earlier members of the kind did in virtue of having the morphology which they share with the new members. So the morphology characteristic of the new members is causally geared to their own proper function. And there is a causal gearing to a further feature as well (Elder 1995, 2007). Ancestor members of such a kind will have produced some effect that made their characteristic morphology get reproduced more widely or more rapidly than alternative historical morphologies, but that reproductive advantage will have depended not just on what the ancestor members themselves did but on the setting in which they did it. There will have been a "historically proper placement" in which ancestor tokens brought about their characteristic effect, and that placement will have played the role of a standing or background condition, enlisted by this characteristic effect in generating reproductive advantage. (An analogy: drug A may outperform drug B as a pain-killer and, if so, will do so in virtue of its microstructure – its "morphology" – but its microstructure confers advantage only against the background condition that human blood can transport A in dissolved form.) Consider, for example, familiar household screwdrivers. New members of the kind get fashioned, on the model of previous members, as a causal consequence of successes effected by the earlier members at fastening certain objects onto others. So fastening objects is their proper function. But these previous successes were possible only because previous members of the kind were environed by suitably slotted screws. So the historically proper placement, for household screwdrivers, includes screws themselves.

In the case of many artifact-kinds, there will be a fairly rich set of essential properties – the "causal gearing" will extend fairly far. For example, screwdrivers must not only be characterized by a certain shape – a "morphology" narrowly so-called – but must be composed of fairly tough and cohesive materials. For they must be capable of turning screws against the resistance of wood or metal in the objects being attached. They must likewise have readily graspable handles. For the performance for which they have been selected is enabling fastening of one object to another *by manual efforts*, human hands too have figured in their "historically proper placement."

In Sect. 4.5, I will consider objections to the effect that the properties characteristic of familiar artifact-kinds are *not* causally geared together as closely as I am saying they are. In order to assess these objections, it will be useful to note, now, that the *way* in which the properties characteristic of an artifact-kind defined by a "proper function" are geared together is different from the *way* in which essential properties are geared together in more familiar examples of natural kinds. Take the kind *salt*, for example, or the kind *gold*. In either case we have a fairly extensive set of essential properties. But all of them (save one) are controlled by a single (essential) microstructural property – by the molecular structure NaCl or by having the electron shells attendant upon having a nucleus that contains 79 protons. So complete is the control that we can be sure that molecular structure NaCl, or atomic number 79, occurs in the members of no other kind. But in the property clusters that characterize artifact-kinds, no one property exercises this degree of control. The same proper function can characterize artifacts having quite different morphologies, as we will see, and artifacts the same in their morphology can belong to distinct lineages, being reproduced on account of historically bringing about different effects – that is, they can differ in proper function. With artifact-kinds, the causal structure is rather this. Any such kind is centrally characterized by a particular morphology, a proper function, and a historically proper placement. No *two* of these properties can occur in the members of a different artifact-kind; the presence of any two will entrain the presence of the third and hence will give us the original artifact-kind all over again. Screwdrivers bearing *that* familiar shape cannot have gotten selected for fastening unless environed by screws; artifacts selected for manual fastening by enlisting screws cannot have gotten selected for unless characterized by roughly that shape; artifacts bearing that shape, selected for because of what they did when employed together with screws, can have gotten selected for only for fastening.

3.4 Does "Thinking Makes It So" Apply to Artifacts' Functions?

The artifacts in each individual kind of artifacts, everyone seems to agree, are essentially characterized by a particular function. I contend that this function should be understood etiologically – as a Millikanian proper function. But will this position help realists with the lower-level worry? This is the worry that, though the status *as* essential of the essential properties of objects is in general mind-independent, still, in the particular case of artifacts, one genuinely essential property is mind bestowed. That property is function. If the function that characterizes this or that kind of artifact is metaphysically grounded on how we act and think with regard to those artifacts, then function appears to be precisely the sort of property that realists cannot afford to regard as real. I now argue that this appearance is misleading – that realists can perfectly well afford to regard artifacts as characterized, indeed as essentially characterized, by a Millikanian proper function.

There *is* a real worry in this general vicinity, I concede. The real worry, for realists, is that the functions characteristic of artifacts will turn out to be "up to us," ours to dictate. That is, the real worry lies in the suggestion that the function of Xs is whatever it is that producers of Xs intend that Xs should do, or that it is whatever the users of Xs intend to bring about by using Xs, or some combination of the two.

But no such suggestion is supported by the position that artifacts of any given kind have a *proper function*, in Millikan's sense. Let us start with the intentions of the producers, and then turn to the intentions of the users. Producers of Xs may intend that Xs should serve uses which do not end up being proper functions of Xs and may fail to intend that Xs serve uses which do end up being their proper functions. The latter case first. What causes a particular pharmaceutical product to have a high sales volume, and thereby explains its profitability and hence its continued presence on the market, may be an "off-label" use for which doctors widely prescribe it and patients commonly ingest it. The producers never intended that the product be used to treat that other malady, but repeated success in treating that other malady is what has caused replication of the drug. Or, a more fanciful example. It is reported that sets of golf clubs sell well to Japanese businessmen who have no prospect of ever using them on a golf course. The producers spent hours designing the clubs to propel golf balls over great distances in straight lines, but their sales – and hence replication – depend, in the Japanese market, entirely on the self-esteem which businessmen gain from viewing the clubs in the closet and displaying them to friends. Now add – this is the fanciful part – that sales *in Japan* are crucial to the marketability of the clubs. Next, the case of unintended proper functions. Nineteenth-century producers of corsets intended that their product should be used to enhance feminine figures. Nonetheless it may well be, as Beth Preston reports, that a large cause of the widespread sale of corsets – and therewith, of the widespread production of corsets – derived from the expression of social status which wearing them provided (Preston 2009, p. 46). If the *main* cause of the widespread replication of corsets was their enabling a declaration of social status, then corsets may well have had a proper function which did not correspond to any use that the producers intended. Another example: the manufacturer of a complex piece of electronic technology may design the product to perform a host of sophisticated tasks which almost no actual purchasers are technologically savvy enough to undertake. The purchasers (who are the users) buy the product because of the dazzlingly complex instruction manual which accompanies it – this being one manifestation of what is sometimes called the "use-plan" (Houkes and Vermaas 2010) – since the manual alone makes them feel sophisticated.[1] But what the producers intended is that the users should *do* the complex things described in the manual and not just glory in the manual itself. The proper function of the device itself consists in only a few simple performances, together with production of a sense of sophistication. For yet another example, consider the recent case of American SUVs (an acronym for "sport utility vehicle"). It is plausible to assume that these were intended by the producers to be used for sporty off-road excursions on which users brought along bulky equipment for camping or climbing or kayaking. But SUVs ended up being sold in great numbers – numbers crucial to their marketability – to urban purchasers whose environment included no off-road destinations whatever.

There are similar deceptions on the side of users. American consumers buy great numbers of elaborate and expensive exercise machines with the intention of losing weight and getting into better physical condition. Most of these machines end up neglected in the basement or posted for resale on eBay. The actual effect produced by these machines, which causally explains their sale and hence their continued replication, is validation of a temporary expectation of fitness, to be achieved in an efficient and nearly sweat-free way. What consumers intend that the machines should produce (one may assume) is *actual* fitness, not merely a transient expectation of it. Or, to take another example suggested by remarks in Preston (2009, p. 47), consider the case of rosary beads. Users employ these in order to bring about spiritual self-improvement and purification. But it would be hard to argue that what causes continued sales and hence replication of rosary beads is actual past successes at contributing to spiritual ends. The more cautious claim would be that their sales depended on consumers' expectations of spiritual improvement, regardless of whether those expectations were veridical.

Now it is true – and this point will be important for my argument in the following section – that all these cases are somewhat atypical. Commonly, perhaps usually, the proper function of artifacts of a given kind will be producing an effect that its producers did intend them to produce and that its users also did expect them to produce. But even in this common sort of case, what is causally central to the continued reproduction of those artifacts – call them "Xs" – is that previous Xs *actually* produced that effect, at least often enough. Indeed there is something misleading here about the casual use of the plural, in both "producers" and "users." That person after person after person has produced an X, or that person after person after person has used an X, often cannot causally be accounted for by adverting merely to intentions harbored in the heads of individuals either group. Often, the groups would not have been there in the first place – there would not have been *many* people playing the role of producers or *many* figuring as users – if Xs had not *in fact* frequently produced the expected effect. The causal efficacy for replication lies in actual past performances, not past intentions or expectations of performance.

[1] Arguably, in such a case the "morphology" of the object that gets replicated should be taken to include the instruction manual itself – just as Dawkins argues that the beaver's dam should be counted as part of the "extended phenotype" of the beaver himself or the bower as part of the bower bird's "extended phenotype" (Dawkins 1999).

The conclusion that I myself am inclined to draw is that even though it is our conscious activities that underlie the characteristic proper functions of artifacts, those proper functions are not "mind-dependent" in a sense that need trouble realists. It does seem important and true that the dependence of proper functions on conscious activities is more than just causal. It does seem to lie in the nature of *artifacts* that they are produced by and for conscious agents and hence that the mechanism of their *re*production involves conscious agents. But no participant in this mechanism of reproduction enjoys armchair epistemic authority over its products. Proper function is no more a case of "thinking makes it so" than is the microstructure of rayon or the weight-distribution characteristic of the Verrazano-Narrows Bridge.

Even so, one philosopher's modus ponens is another philosopher's modus tollens. Philosophers inclined to maintain that the functions of artifacts are mind-dependent features, of a sort that realists cannot afford to recognize, might elect to maintain that the microstructure of rayon, or the shape of the Verrazano-Narrows Bridge, is likewise a realism-unfriendly mind-dependent feature. I need to offer something better than a blasé shrug in order to establish that realism, in and of itself, is compatible with the admission that there obtain in the world features which we conscious agents causally instill in their bearers.

Here is my suggestion. What is central to realism is rejection of the idea that "thinking makes it so." Realism must maintain, then, that for any real property *p*, there is no conscious individual *i* and no conscious group *g* such that *i*'s thinking (or *g*'s thinking) that *p* obtains in such-and-such property bearers is *constitutive* of *p*'s obtaining in those property bearers. Similarly, for no real kind *K* is there some individual *i* or some group *g* such that *i*'s thinking (or *g*'s thinking) that *K*s persist across such-and-such changes, or are to be found in such-and-such locations, is *constitutive* of *K*s existing in those contexts. Realism, I suggest, is a negative ontological claim about what grounds the existences of the world's objects and the possession, by those objects, of their properties. But realism can equally – and perhaps more usefully – be conceived as an epistemological claim about epistemic privilege. For, given the plausible assumption that if *i* or *g* are entertaining certain thoughts, *i* or *g* will *know that* they are entertaining these thoughts, what realism holds is that for no real property *p* or kind *K* is there some party – some individual or some group – such that that party, just by virtue of *being* the party it is, enjoys effortless armchair authority as to where *p* obtains or where *K*s exist. So realists can afford to allow that there are kinds of objects each essentially characterized by some proper function – that is, artifact-kinds – so long as realists can plausibly say that neither producers, nor users, nor anyone else enjoys effortless armchair authority concerning what those proper functions are. But precisely that claim is plausible, as the reflections above show.

3.5 "Causal Gearing" and Capitalist Artifacts

The strongest reasons for doubting whether realists can affirm artifacts stem from the premise that artifacts of each kind are essentially characterized by some function. That premise is threatful just because it seems clear that the functions of artifacts are metaphysically grounded on the ways that conscious beings act and think with regard to them. But realists can afford to agree with both premises, I have argued, by contending that the functions characteristic of each artifact-kind are historically constituted proper functions. There is a cost to this contention: what realists identify as "the function" of artifacts in this or that kind will sometimes diverge from what common sense says those artifacts "are for," what they "are supposed to do." But there is a countervailing benefit. These proper functions can be said to be causally "geared together," with other properties that seem essential to the artifacts in each kind, in just the way required for them to qualify as mind-independently essential. They are geared together, or so I have argued, with both the morphology and the historically proper placement characteristic of each artifact-kind.

But is it really true that causal gearing connects proper function with such other properties characteristic of artifact-kinds? Amie Thomasson has argued that precisely the sorts of arguments that I offer here fail to establish that conclusion (Thomasson 2007b). Consider two kinds of artifacts that Thomasson offers as examples, namely paper clips or corkscrews. Paper clips and corkscrews have come to be made in many different shapes, while all retaining the same proper function. So shape – that is, morphology – seems to float free from the other properties that my realist position treats as essential to these artifact-kinds. Nor can I consistently claim that these are isolated examples. Above, in arguing that realists can afford to recognize Millikanian proper functions, I repeatedly invoked the idea that continued replication of a given artifact-kind requires marketability and hence profitability. I thereby situated the artifact-kinds I was discussing in specifically capitalist relations of production. But capitalism virtually guarantees a proliferation of new designs for products that serve familiar functions, functions for which consumer demand has been established. The gearing that a realist must believe in seemingly comes loose across a wide range of artifacts.

My basic response is that (as I said in Sect. 4.3) the *way* that essential properties are geared together, in the natures characteristic of artifact-kinds, is different from the *way* essential properties are geared together in the natures of well-worn examples of natural kinds such as *water* and *gold*. Typically, an *individual* property incorporated in the nature of an artifact-kind can be present among the members of different artifact-kinds. What cannot happen is that *two* individual properties, incorporated in the basic triad "proper function, morphology, and historically proper placement," should occur among members of a different artifact-kind. The proper function of corkscrews having the morphology of the typical "pull up" design does indeed occur also in corkscrews having wings which, when depressed, effect extraction of the cork. But, I contend, the historically proper placement for "pull up" corkscrews differs from the historically proper placement for "winged" corkscrews. Despite the fact that we use the sortal "corkscrews" for items characterized by either morphology, the items actually belong to distinct artifact-kinds.

This line of response may initially seem unpromising. It may seem implausible and ad hoc to say that when a new design is introduced, for what we speak of as "the same sort of device," a different historically proper placement is brought into play. Actually it is fairly easy – as I shall illustrate below – to identify, in such cases, *minor* differences in historically proper placement. But to justify the idea that the historically proper placement differs *enough* to warrant talk of a different *kind* of artifact, I need to say something about a doubling of "proper function" that is characteristic of artifacts produced under capitalist relations of production. Fortunately, what I have to say is true.

What it is that gets artifacts bearing a certain shape or design to be produced over and over, in a market economy operating on basically capitalist lines, is that previous tokens of that type have enabled owners of the means of production to make a profit – indeed a better profit than they would have made, using the means of production at their disposal in any readily available alternative way. Over the long run, under capitalism, this sort of relative profitability is both causally necessary and causally sufficient for continued replication of a given design.

What then is *the* proper function, under capitalism, of the artifacts in a given historical lineage – the individual artifacts in which the design of some prototype gets copied over and over? If proper function is a matter of what has caused replication of a design, it follows that the *first* answer about proper function for capitalist artifacts is the same for every historical family of artifacts: their proper function is yielding, for the manufacturer, this sort of relatively optimal profit. But of course that cannot, in the usual case, be the whole story on proper function. Usually, artifacts in any historical lineage must do something that is of use or of interest for the consumer, if the consumer is to judge that in purchasing them he is serving his own interests and not just those of the manufacturer. Moreover, this useful performance must be fairly fixed and learnable, for consumers to have an idea of why they are buying *that* product, and must actually be carried out by the product *often enough* for consumers to believe that their purpose will be served. Capitalist artifacts of any one copied kind must then have a *second* proper function as well, a fixed and learnable performance that is of interest to consumers. Adapting terminology of Millikan's, we can say that the *focused proper function* of the capitalist artifacts in any one kind is yielding relatively optimal profit, and this the artifacts do as a causal consequence of their serving a *proximal proper function* of interest to consumers (Millikan 1984, pp. 33–38). *Both* proper functions are constitutive of the very nature of the artifacts in any one copied kind. (To put the same point using Marx's terminology, capitalist artifacts are created and shaped by the demand to maximize exchange value, and this they generally can do only by having identifiable use value.)

This doubling of proper function makes it possible to argue that there really is "causal gearing" among the features comprised in the nature of a particular kind of capitalist artifact, even though capitalism ceaselessly calls forth new designs (or new morphologies) in artifacts that serve a familiar use. If morphology could vary freely while all aspects of *both* proper function *and* historically proper placement stayed fixed, we would indeed have a violation of "causal gearing." But when we note that there is a double proper function for capitalist artifacts, we can see that a new morphology may enlist a new historical placement for the function of profit maximization, even while it enlists the same old placement for the function that attracts consumers. Here is an illustration from the particular case of corkscrews. When corkscrews of the easier-to-use "winged" design were developed, they proved profitable, despite the cost of developing the new equipment required to manufacture them, because they appealed to consumers who wanted to open wine bottles, but who were too weak or too clumsy – don't be offended, dear reader, as I include myself in this group – to find the old "pull upwards" design convenient. These new corkscrews then served their *focused* proper function, i.e., profit yielding, because they enlisted the historically proper placement of being marketed to weak and clumsy wine lovers. So there was a change not *just* in design. There was a change in historically proper placement for performance of the *focused* proper function. The *proximal* proper function stayed the same – it was still the opening of wine bottles. The historically proper placement for performance of that proximal proper function stayed largely the same – the corks which the new corkscrews served to remove had the same shape and composition as before and were emplaced in wine bottles having the same shape as before. But there was a change in the other historically proper placement, i.e., in the consumer market, and hence there occurred a combination of essential properties not found among corkscrews of the older, "pull upwards" kind.

I suggest that a corkscrew-like story may hold true even for the case of paper clips. Paper clips of the older "trombone" shape are a bit difficult to apply to thick stacks of papers. The papers in the stack get clamped together at just a single point, and the friction involved in pushing the paper clip over the stack is concentrated at that single point. Paper clips of the two-pointed "butterfly" shape intersect a stack at two different points, and the friction can more easily be overcome by lightly rocking the clip back and forth. In any case, *something* has to have conferred a market advantage for the "butterfly" shape, at least among some consumers – otherwise the development of that design would merely have hindered profitability – and I am suggesting that this greater ease in application is what did it. But then there has to have been a segment of the market that would respond to this greater ease. This segment constituted the historically proper placement for performance of the *focused* proper function – that is, profit yielding – of paper clips having the "butterfly" shape.

Or consider a sort of case that Thomasson does not mention, but certainly *might* have: what of the case in which artifacts of some outmoded design continue to be manufactured, and continue to serve the focused proper function of yielding profit, because of performing some *new* useful job for consumers? The shape stays the same, and the *proximal* proper function has altered; so, is this a case in which proximal proper function comes loose and fails to be geared to other kind-characterizing properties? I am thinking of such cases as that of the huge industrial spools once used for thread, and now sold as coffee tables, or of old-fashioned non-electric irons no longer used for ironing clothes but instead used as nostalgic paperweights. But here too, I contend, we have not merely a change in proximal proper function, but a causally connected change in historically placement – this time, in the historical placement proper for performance of the *proximal* proper function itself. The giant spools get sold because people have coffee cups and magazines that they want to have supported; the non-electric irons get sold because people have stacks of bills that, for sad reasons, they do not want to have dispersed by the wind.

The thesis of "causal gearing," then, has greater resilience than one might at first suppose. Even so, one must admit that the phenomenon which the thesis captures is coarse grained. Suppose that manufacturers of a familiar kind of artifact could introduce a trivial alteration in its shape without incurring measurable production costs. Then the trivial alteration might actually occur, even without enlisting demand among more consumers or different consumers. It might spread on account of something more like genetic drift than like natural selection. Only a *roughly defined* morphology, then, would be geared to just *that* familiar use and *those* particular consumers.

Let me close by addressing a way in which the doubling of proper function, for capitalist artifacts, may seem threatful for realism in particular. The focused proper function of such artifacts is that of profit maximization, as described above. But just about everyone familiar with capitalism *knows* this! The worry, then, is that producers, or perhaps consumers, are in a position to judge, with effortless armchair authority, what the focused proper function is of the artifacts in any one kind.

My reply is that the focused proper function of capitalist artifacts is not mind-dependent in any way that need embarrass realists. It is an objective economic fact that a particular artifact-kind will continue to get replicated, under capitalism, only so long as sales of its instances return a relatively optimal rate of profit. Many individuals and many groups rightly judge this to be the case, but no one of these parties is right in this judgment just in virtue of *being* the party that it is. Producers, in virtue of being producers, may intend that their product return a relatively optimal rate of profit. But even if – as we may plausibly assume – producers will know *that* that is what they intend their product to bring about, it does not follow that producers are, by virtue of being the producers, bound to be right about that relatively optimal return of profit *will* be what their product brings about. For their product may cease to exist: the market for the product may collapse, as happened with the recent near collapse of the American automotive industry. To be sure, producers may know that *if anything* causes continued replication of their product, it will be the yielding of relatively optimal profit. But they will know this only by being perceptive students of capitalism in general – not in virtue of their armchair authority as to what their own *intentions* are. To see this, consider a case – not entirely fanciful – in which an ecologically minded producer elects to produce some "green" product even though its high cost of production, and weak sales, will yield a substandard rate of profit. What happens to such producers, at least in general and over the long run, is that investors withdraw their capital, and invest it elsewhere, where a higher return on their investment is possible. The producers intend that their product should yield a modest rate of profit, but their intention fails, since their product ceases to exist. So whether producers intend their product to produce optimal profit, or intend it to return only modest profit, neither intention causally determines what rate of profit that product – if indeed there continues to be such a thing as "that product" – returns.

3.6 Cultural Artifacts and "Thinking Makes It So"

Now for a different category of artifacts. Consider things such as gender roles and codes of etiquette, canons of style in clothing and personal ornamentation, and rules for correct word choice and usage. These seem to be artifacts of a culture – cultural artifacts. But they raise, for the realist, a sharp worry about "thinking makes it so." This worry does not, as in the previous section, concern the functions of cultural artifacts. Maybe only Marxists know about *that*. This worry concerns the content or shape of such cultural artifacts – their "morphology," in a transposed sense. That is, it appears that our epistemic position with regard to the content of our canons of style and beauty, of our gender roles and of our codes of etiquette, is a matter of "effortless armchair authority" – in just that sense in which realists must deny that anyone has such authority with respect to the nature or whereabouts of anything real.

In a moment I will say more about the effortless authority that we do appear to enjoy here. First let me say why, in explaining the threat to realism that such authority appears to pose, it is crucial to use "our" and "we." Gender roles (if there are such things) specify the differences between the ways it is appropriate for men to behave and the ways it is appropriate for women to behave; codes of etiquette fix what is polite and what is rude; canons of style determine what is chic and what is frumpy, what is beautiful and what is homely. But everyone of any sophistication understands that the facts thus fixed are facts that obtain only relative to a particular group. There is no fact of the matter as to what is polite and what is rude *tout court* or what is polite and what is rude for everyone whatever. There is only a fact of the matter as to what is polite or rude "for us": whomever or whatever "us" designates, there will be other groups for whom what is rude for us is not invariably rude, nor what is polite, and likewise for what is chic or what is gender appropriate.

Now for the apparent threat to realism, it seems to be a serious truth that certain sorts of behavior *are* polite or rude for us if and only if we *take them to be* polite or rude; that to the extent that we any longer suppose there are differences between the ways women can appropriately behave and the ways men can appropriately behave, the content that we *assign* to these ways fixes, and exhausts, what our gender roles *are*; and that what we take to be the difference between a *mot juste* and a maladroit way of putting a point is constitutive of good usage for us. Whatever "we" think, regarding politeness and rudeness, beauty and homeliness, and correct word usage and incorrect word usage, fixes the facts – that is, the facts-for-us – concerning those features. All the areas I have mentioned seem to be domains in which "thinking" – that is, "our thinking" – "makes it so."

But a question still remains: just who or what might enjoy effortless armchair authority concerning the facts-for-us in these various domains? Well, *we* do, I have just said. But does "we" designate a unitary Group Mind, a unitary "collective subject"? The idea that there might be such a thing as a Group Mind seems suspect, though I will return to it later. The more modest reading seems to be that individual ordinary human beings enjoy effortless armchair authority in these areas. Perhaps we should suppose that every individual comprised within "us" enjoys such authority. Or perhaps we should recognize a "cultural division of labor" and rule that we individuals who are comprised in "us" enjoy authority vicariously, by virtue of the "experts" among us who themselves – in their own right, and *not* vicariously – enjoy epistemic authority with respect to these matters. The latter answer is surely the more promising. It seems virtually a truism that "some of us" may have bad taste, or may have a boorish ignorance of etiquette, or be clueless as to the differences between the ways it is appropriate for men to behave and for women to behave.

But are our experts really in a better position? The facts as to which they would be (non-vicariously) authoritative are, as we noted, relativized facts. They are the facts as to what is polite-for-us, frumpy-for-us, and well-said-for-us. Can a person be said to know such a fact without knowing whom "us" encompasses? It obviously is *not* required that our experts be able to list, by name, all the individual human beings for whom they enjoy expert status. But it does seem plausible that to know what it is polite for us or chic for us or gender appropriate for us, the experts must have some general idea as to where this "us" is located – in what countries, what geographical areas, which linguistic communities – and how far it extends back in history.

This will not be easy to know. The groups whose conscious activities underlie the existence of such things as gender roles, codes of etiquette, and canons of style are typically large, diffuse, and have vague boundaries. Indeed those boundaries – such as they are – would seem to be fixed by causal factors concerning which no one enjoys epistemic privilege. Where *something* causes the individuals in some plurality to act and think in distinctive common ways, there do we have the makings for such a group. This "something" may have to do with the social emplacement of these individuals, or with their personal attributes, or merely with overlapping strands of personal acquaintance. It may matter crucially that the individuals between them play some distinctive cultural or political (or economic or historical) role or are distinctively affected by some cultural or political process. But in any case, concerning who is and is not included in typical and interesting social groups – those which collectively bring into existence such cultural artifacts as we are considering – it will not be the case that "thinking makes it so," for any ordinary human thinker.

I conclude that, because the facts that articulate our gender roles and codes of etiquette and other cultural equipment are relativized facts – facts that obtain only for a vaguely, but causally, bounded group – *no* human beings are in an epistemic position of armchair authority with respect to them.

Now back to the idea of a Group Mind, a "collective subject." Margaret Gilbert has for years defended an ontologically modest version of this idea (Gilbert 1987, 2004). There may be truths as to what we collectively believe, she has shown, that do not reduce to truths about what we individually believe. But *that* we collectively believe these things nevertheless amounts to facts about ordinary, individual human beings. It amounts to the fact that we, as individual human beings, jointly commit ourselves to acting, collectively, as if collectively we have certain beliefs. Yet this modest version of a Group Mind puts no actual mind in the epistemically privileged position of knowing, with armchair authority, just who is participating in this "collective subject." So even if our collective beliefs fix the facts-for-us about gender or etiquette or fashion, the problem recurs about knowing just whom "us" encompasses.

But *does* anyone really need to know whom "us" encompasses, in order for it to be the case that realism has serious trouble with "cultural artifacts"? Realism, I have said, is fundamentally an *ontological* principle – a principle about what does *not* underlie the obtaining of properties or the existence of objects. Only secondarily, as

I represented the situation, is it an *epistemological* position – one that denies armchair epistemic authority. Might one dispute the linkage? Might one say that it is after all *our* ways of thinking that fix the facts-for-us about etiquette, fashion, and gender appropriateness, even though *we* are so sprawling and vaguely boundaried a plurality that no one knows whom *we* encompasses? But I question whether it makes sense to say that there is such a thing as *our* thinking even though *we* are so sprawling and decentralized a plurality that no one is in a position to know, authoritatively, just what *we* think. It seems less mysterious to say there are "experts" who know perfectly well what they think and who rule by relations of power the nature and reach of which no one authoritatively grasps.

References

Dawkins, R. (1999). *The extended phenotype*. Oxford: Oxford University Press.

Elder, C. (1992). An epistemological defense of realism about necessity. *Philosophical Quarterly, 42*, 317–336.

Elder, C. (1995). A different kind of natural kind. *Australasian Journal of Philosophy, 73*, 516–531.

Elder, C. (2004). *Real natures and familiar objects*. Cambridge, MA: MIT Press.

Elder, C. (2007). On the place of artifacts in ontology. In E. Margolis & S. Laurence (Eds.), *Creations of the mind: Theories of artifacts and their representation* (pp. 33–51). Oxford: Oxford University Press.

Elder, C. (2011). *Familiar objects and their shadows*. Cambridge: Cambridge University Press.

Findlay, J. N. (1963). *Meinong's theory of objects and values*. Oxford: Oxford University Press.

Gilbert, M. (1987). Modelling collective belief. *Synthese, 73*, 185–201.

Gilbert, M. (2004). Collective epistemology. *Episteme, 1*, 95–107.

Houkes, W., & Vermaas, P. E. (2010). *Technical functions: On the use and design of artifacts*. Dordrecht: Springer.

Millikan, R. (1984). *Language, thought, and other biological categories*. Cambridge, MA: MIT Press.

Millikan, R. (1993). *White Queen psychology and other essays for Alice*. Cambridge, MA: MIT Press.

Preston, B. (2009). Biological and cultural proper functions in comparative perspective. In U. Krohs & P. Kroes (Eds.), *Functions in biological and artificial worlds* (pp. 37–50). Cambridge, MA: MIT Press.

Sidelle, A. (1989). *Necessity, essence, and individuation*. Ithaca: Cornell University Press.

Thomasson, A. (2007a). *Ordinary objects*. New York: Oxford University Press.

Thomasson, A. (2007b). Critical notice of *Real natures and familiar objects*, Crawford L. Elder. *Philosophy and Phenomenological Research, 27*, 518–523.

References

Chapter 4
Public Artifacts, Intentions, and Norms

Amie L. Thomasson

Abstract Artifacts are often said to be things intentionally created to serve a certain function, where function plays the dominant role in classifying artifacts into artifactual kinds. Here I argue, however, that artifacts need not have intended functions and that even when they do, that does not always play a core role in artifactual classification. Artifacts, I argue, must have intended *features*, but these may include not only functional but also structural, perceptible, or even receptive and normative features regarding how the object is to be regarded, used, or treated. Indeed, I argue that members of *public* artifact kinds depend on the existence of public norms of treatment. Recognizing the role of receptive and normative features in public artifact kinds enables us to provide a better account of artifact categorization, solve old puzzles about exaptation and minimal creation, and provide a better understanding of the significance of artifacts in our lives and in the social sciences.

Keywords Artifact • Artifact kind • Intended function • Normative features • Public norms • Works of art

4.1 Introduction

"Artifacts are objects intentionally made to serve a given purpose," writes Lynne Baker (2008), echoing standard dictionary definitions of "artifact" and expressing a commonly held view. The popular idea that artifacts are objects created with a

A.L. Thomasson (✉)
Department of Philosophy, University of Miami,
414 Majorca Ave, Coral Gables, FL 33134, USA
e-mail: thomasson@miami.edu

M. Franssen et al. (eds.), *Artefact Kinds: Ontology and the Human-Made World*,
Synthese Library 365, DOI 10.1007/978-3-319-00801-1_4,

certain intended function leads to three commonly held views about artifacts.[1] First is the idea that artifacts are *functional* objects and that for something to be an artifact, it must have an intended function. Second, given the centrality of the intended functions to our artifact concepts, it has also been widely supposed that not only must an artifact have an intended function, but also that its intended function determines (or plays a central role in determining) what *kind of* artifact it is. Hilary Kornblith writes "At least for the most part, it seems that what makes two artifacts members of the same kind is that they perform the same function" (1980, p. 112).[2] Thus, we classify vegetable peelers and can openers differently, since each has a different intended function. Third, artifacts are *intentional products,* things intentionally created by humans (or perhaps other intelligent creatures), and thus are mind-dependent objects in the sense that it is a conceptual truth that an artifact comes into existence only if it is intentionally made. Thus, artifacts are often thought to be dependent on the individual intentions of their maker, but are not generally thought to depend on any other intentional states.

While this is a natural way of understanding artifacts, I will argue that none of these claims is quite right. First, I will argue, something may be an artifact without having an intended function. Second, even where artifacts have an intended function, having that intended function is not always necessary or sufficient for classification in the relevant artifact kind. While artifacts must have certain *intended features*, I will argue, these need not be limited to (nor even include) an intended *function*—they may also include structural, perceptible, or other features, which may also (or instead) serve as essential intended features for members of the kind.

After making those arguments rather briefly, I move on to focus on those artifacts that are members of familiar, recognized, public artifactual kinds: things like forks, computers, cars, statues, clothes, and the like. For short, I will refer to these as "public artifacts." I will leave out of discussion useful objects that an individual might privately create for some novel need. One might argue over whether or not the latter are properly called "artifacts"—one could apply the term to them and insist that what I am calling "public artifacts" are a mere subclass of artifacts, or one might decide to restrict the term "artifact" to public artifacts and call the rest "tools" or "use-objects."[3] The very term "artifact" is itself used quite loosely, and in many different ways, so there may be no single characterization of what is essential to artifacts that fits best; whether or not we include the latter among artifacts is largely a verbal issue which I have no special interest in adjudicating.

What I do have interest in is turning attention to these public artifacts. For even if one insists that these are merely a subclass of artifacts, they are those artifacts that

---

[1] Though I do not mean to be attributing all of these views to Baker herself.

[2] Kornblith himself does not make it clear, however, what sense of function (intended function, actual functional capacities, proper function, or some other notion) he has in mind.

[3] This use of "tool" as opposed to "artifact" is related to Dipert's (1993, pp. 27–29). "Use-object" is Husserl's term for an object useful for some individual's purposes, but he does not restrict the term to artifacts – natural objects (such as lumps of coal) may also be use-objects in his sense (1989, p. 197).

play the most central role in our common life and are those of primary interest to archeologists, historians, anthropologists, museum curators, and the like. These public artifacts, I will argue, are of particular interest since studying them calls attention to another range of properties that may serve as intended properties of artifacts, beyond the functional, structural, and perceptible, and which may be criterial for membership in a given artifactual kind. These are what might be broadly construed as receptive and normative features, involving how the object created is to be regarded, used, treated, or behaved in regard to (and by whom, in what context). As a result of the fact that such receptive and normative features are typically among the essential features of members of public artifact kinds, I will argue, another interesting result follows: While all artifacts are indeed mind dependent, public artifacts do not depend merely on the *individual* intentions of their makers; they also depend on public norms.

Emphasizing the role of recognitional and normative features among the defining features of public artifact kinds can help us resolve puzzles about exaptation and minimal creation. More importantly, it provides an understanding of artifacts as objects that are not merely useful, but infused with *significance* for our lives and actions. This enables us to better capture our experience of dealing with artifacts and make clear the relevance of artifacts for social science and for our daily lives and activities. Finally, it calls attention to the inherent normativity in artifacts, bringing to the fore commonalities between language and other artifacts.

4.2 Artifacts and Intended Function

Artifacts are often characterized as objects created to serve a certain intended function, and certainly many of our paradigmatic artifacts have an intended function.[4] Two questions may be raised here: First, must artifacts have an intended function? Second, is having a particular intended function essential to classification in an artifactual kind? I will argue that, despite the core role intended function plays in many artifacts and artifact kinds, the answer to both questions is "no."

There are well-known problems with the view that all artifacts have an intended function. Such things as doodles and idly produced paper clip sculptures do seem like artifacts, though they may lack any intended function. More broadly and importantly, works of art often seem to have no intended function: It seems that an artist may create a painting or symphony without intending it to have any function whatsoever—but that in so doing, she has nonetheless made an

[4] There has, of course, been substantive debate about whether the functions of artifacts should be understood as their intended functions, actual causal capacities, proper functions (in something like Millikan's (1984) sense of functions acquired in virtue of their history of production), or in some other way. I do not intend to enter that debate here, as I am here more narrowly focused on arguing against the requirement that artifacts have an intended function. For a careful and thorough discussion of the debate about artifact functions, see Preston (2009).

artifact. So it seems that we should allow that there may be artifacts that lack an intended function.

One could move straight from there to argue that if there are artifacts without an intended function and if every artifact belongs to some or other artifact kind, there must be artifactual kinds for which having a particular intended function is not essential. One could, perhaps, question whether every artifact belongs to some artifact kind and doubt the argument's soundness on those grounds. But looking at particular artifact kinds also gives us reason to reject the view that, for all artifact kinds, having a particular intended function is essential to classification in that kind. We might naturally look again to the case of works of art. Even where works of art have an intended function (e.g., where a film has the intended function of persuading viewers of a political point), possessing that function is neither necessary nor sufficient for being a film. There may be sub-kinds of art (e.g., altarpieces) that all share some function, but the broader art kinds we commonly traffic in, such as paintings and symphonies, clearly have no essential intended function—paintings have been created with all sorts of purposes in mind, from decorative to documentary, from self-expressive to religious to political, and yet all apparently belong to the kind "painting." Moreover, as Paul Bloom has aptly pointed out, even members of artifact kinds that are associated with a function—artifacts like boats or chairs—need not be intended by their producers to serve that function. Something may be a boat or a chair even if its maker intends that it never serves the intended functions of boats or chairs, but be merely "for show" (1996, pp. 5–6). Thus, it seems clear that, at least for many artifact kinds, having a particular intended function is not necessary for membership in those kinds nor is it sufficient. Many other artifactual kinds have defining structural or perceptible features as well as functional features: For something to be a double-breasted suit or a Corinthian column, it is not enough that it shares the intended functions of providing (socially respectable) clothing or independent support of a roof; it must also have certain characteristic shape or decorative features.

This all seems to provide good reason to reject the idea that artifacts must be objects with a certain intended function, and along with it to reject the idea that intended function invariably plays a core role in determining artifact kind membership. In place of the idea that artifacts must have an intended function, we should generalize the idea, treating artifacts as things that are intentionally made and which have at least *some* intended features—which may or may not include an intended function. Risto Hilpinen expresses this idea in his broader definition of an artifact, as something "intentionally produced by an agent under some description of the object" (1992, pp. 59–60) (at least one of which must be a sortal description) and having some intended properties. Hilpinen also requires that the object produced actually exhibits some of the intended properties (including the sortal property), building in a success condition for the production of artifacts: An artifact may not be produced by intentions *alone;* as Hilpinen puts it, "an agent produces a genuine artifact only if his activity is *successful* in some respect and to some degree" (1993, p. 160).

Similarly, I have argued elsewhere that we should adopt a more general view of what (intended) features figure in classifying artifacts: that something is a member of an essentially artifactual kind[5] K only if it is the product of a largely successfully executed intention to make a K, where the maker must have a substantive concept of the nature of Ks that largely matches that of some prior K makers (if any) and intend to realize that concept by making an object with K-relevant features (2003, pp. 599–600). This account allows K-relevant intended features to vary from kind to kind: They may of course include intended function, but also may include shape or structural properties (often among the properties definitive of membership in clothing kinds and architectural kinds such as bell-bottoms or an A-frame cottage), taste properties (e.g., for culinary kinds, if (say) candy must have sweetness as an intended property), and so on.

Recognizing that the intended properties of artifacts that are relevant to membership in the relevant artifactual kind need not be, and typically are not, merely functional properties, but may also include structural properties, sensory properties (flavor, color, sound), aesthetic properties, and so on, is also important to offering a more appropriate understanding of our artifactual categorization. For it enables us to group paintings together even when they do not serve a common function, enables us to classify those chairs and boats intended for "show" with their working cousins, and enables us to account for finer-grained distinctions in artifact kinds defined as much by intended structural or perceptible features as by intended function.

4.3 Intended Recognitional and Normative Features

All of this I have discussed elsewhere. What I want to do here is to call attention to an additional, often unnoticed, range of intended features that may be relevant to membership in many of our familiar, extant, artifact kinds: those intended properties that involve not just perceptible or functional features of the object created, but rather intended ways in which the object is to be regarded, used, or treated. To have a general term, I will call these "receptive" features.

The importance of receptive features has occasionally been pointed out before. Randall Dipert, for example, draws a three-way distinction between instruments (objects intentionally used to serve some purpose, but which may be natural); tools (instruments that are intentionally modified to serve some purpose [or intentionally left alone]); and artifacts, where "an artifact is an intentionally modified tool whose modified properties were intended by the agent to be recognized by an agent at a later time as having been intentionally altered for that, or some other, use" (1993, pp. 29–30). So, for example, as Dipert notes, chairs are objects intended to be recognized as having been intentionally made as a seating device, and concludes that "artifacts are, unlike tools, distinctively 'social'" (1993, pp. 30–31) and have a "communicative purpose," since they are made with the intention to bring about a belief in another agent (1993, p. 102).

[5] Where an essentially artifactual kind is a kind that necessarily has in its extension all and only artifacts – considered as intended products of human action.

One must be careful, however, in characterizing what kind of recognition is intended. It does not seem quite right to say that to be a teapot or a poem or a ring, something must be created with the intention that it be (actually) *recognized* (by anyone other than the maker) *as* a teapot, poem, or ring, since one may in principle make an entity of any of these kinds intending it to be kept completely private and never seen by anyone except its maker. So we should at least modify the view to say that they must be intended to be *recognizable* as members of that kind. This intended recognizability, moreover, need not be completely general, but may be with respect to a certain intended audience. Hilpinen makes this point noting that "a mechanical shark used in making an adventure film is an artifact, but its authors do not wish the audience to recognize it as such, on the contrary; the condition of recognizability presumably applies only the persons who are using it in the making of the film" (2008). Dipert himself notes that spy objects (like a microphone disguised as a martini olive) may be intended *not to be* recognized—at least by most people (1993, p. 31)—but suggests that we may still consider it an artifact if we count the spy himself as the one intended to recognize it. If, on the other hand (as seems more likely in spy circles), the object's kind *is* intended to be recognizable *by certain intended individuals* (spies for the relevant party)—perhaps by including it with the proper instruction manuals as it is passed on, etc., then the recognizability condition holds, as long as we express it as the idea that artifacts must be intended to be *recognizable* by a certain *intended audience*. Even standard artifacts like chopsticks are intended to be recognized as chopsticks only by a certain intended audience—i.e., those properly enculturated into chopstick use practices, so the idea that intended recognizability is relative to an intended audience is quite general, even if the size of the intended audience varies.

Intended recognizability (by an intended audience) may, however, still seem too strong a condition to cover absolutely all artifacts. One can perhaps imagine cases in which someone intentionally creates something that will not be recognizable even by herself (e.g., I intend to make something that looks like a rock to help fill in a planter I have and intend that it matches the others so well that neither I nor anyone else will be able to recognize which is the artifact).[6] As mentioned at the outset, I do not mean to deny that the term "artifact" may, in some contexts, be used quite broadly, to also apply to private tools, use-objects, or other private creations. So we can certainly count this as an artifact in the broad sense. But what I aim to focus attention on here is members of public artifactual kinds, and we can deny that the "rock" created is a member of a public artifactual kind, even if it is counted among artifacts. With this in mind, we can also say that if the spy gear above is the spy's private personal creation, it is not a member of a "public artifact kind" as I am using the term here. Thus, so far, it seems that a recognizability criterion is defensible if we express it as the idea that members of *public* artifact kinds must be intended to be *recognizable* as artifacts of that kind by a certain *intended audience*.

[6] Thanks to Simon Evnine for this point.

But the intended features need not be merely recognitional, they may also involve other ways in which the creation is to be considered, regarded, or treated. Jerrold Levinson holds that what unites artifacts as *works of art* is not any intended structural, aesthetic, or perceptible properties, but rather intentions about how the products are to be regarded. The basic idea of his intentional/historical theory of art is that "something is art in virtue of being governed by certain intentions with an essential historical, or backward-looking, content" (2007, p. 74). More specifically, "an artwork is something that has been intended by someone for regard or treatment in some overall way that some earlier or pre-existing artwork or artworks are or were correctly regarded or treated" (2007, p. 74). Thus, in this case, the idea is that the relevant intended features that unify objects into the kind "work of art" are *purely* regard-based features: intentions about how the object is to be considered or treated. Whether or not one accepts this as a final view about how to define "art," it certainly seems right that intended features about how the product is to be regarded, treated, or behaved towards (and by whom in what context) may play a far more central role in the classification of artifacts into *art* kinds than intended functional features do.

The most developed accounts of the sense in which intended receptive features may be essential to membership in (public extant) artifact kinds were offered long ago, however, in the phenomenological tradition. So, for example, Roman Ingarden notes that the difference between a piece of cloth and a flag lies largely in the different norms of treatment each is subjected to:

> With a piece of cloth, for example, we clean pots. To the flag we render military honors; we preserve it, often for centuries, as a remembrance, even though the cloth of the flag is badly damaged and without any value. (1989, p. 260)

Similarly, Ingarden suggests that one constitutive difference between different kinds of building, such as a church and a theater, lies in the different norms of comportment demanded of each: "This manner of comportment [required for churches] conforms to the views dominant in the pertinent religious community, and under different circumstances and in different cultural buildings, such as a theater or a club, would be inappropriate and even ridiculous" (1989, p. 260). Different behavioral norms may be placed on different sorts of people (e.g., believers versus nonbelievers, or, we might add, men versus women, adults versus children, laypersons versus various officials of the church)—so the norms imposed may not be uniform norms of "how this building is to be treated or regarded," but also *by whom* and *in what circumstances*. Moreover, as Ingarden also insightfully notes (1989, p. 261), many of the intended structural features of churches in fact are present not to serve any (other) practical function (people might just as well gather in a theater and hear the priest even better there) but to serve the *recognitional* function of making it known as a church of a certain type. The intention that the object be *recognizable* (by an intended audience) as a member of the kind in turn serves the further purpose of enabling the intended audience to recognize the object as *to be treated* in the appropriate ways—as subject to the relevant norms.

In short what seems most basic in many cases is the intention that the creation be *subject to certain norms*, in the sense that it be *recognizable* as something that is *to be treated, used, or regarded,* in some ways rather than others (in some contexts, by some individuals…). It is the intended normative features (that the object be subject to certain norms) that drive the intended recognitional features (noted by Dipert) as well as many intended structural features.

The idea that there may be what we might call "constitutive norms" of treatment for culturally significant objects, including public artifacts, receives perhaps its earliest development in Heidegger.[7] Heidegger, of course, was not concerned with artifacts as such, but rather with the wider class of objects "ready-to-hand" (which include natural objects with a standard role in our way of life).[8] These objects ready-to-hand, on his view, are distinguished by having features such as situatedness in a range of equipment or "equipmental contexture" in which they *belong*, having certain intended users ("for whom" they are), certain goals or purposes "towards which" they are *to be* used, and certain norms of how they are *to be* used (usability). Objects ready-to-hand, we might say, are enmeshed in norms regarding their appropriate context/placement, users, use goals, and use practices. Thus, part of what it is, on this view, to be a baptismal font is to be something to be used *by* certain people (priests on the outside, babies on the inside), *for* certain purposes (to initiate the child into the church), *in a certain way* (by sprinkling water or immersing the child), in certain contexts (near the nave of a church, as part of a baptismal ceremony), etc.

Notice, though, that there is a difference between Heidegger's treatment of the ready-to-hand as something (actually) subject to certain norms of treatment, and the suggestion above that members of extant public artifact kinds are the products of largely successful intentions, among them that they are *to be subject to certain norms of treatment.* Heidegger, of course, was considering the wider category of entities ready-to-hand; it is entirely plausible that purely natural objects be subject to norms of use (that round stones, in this culture, are *to be collected and placed around the fire,* that cows are *to be revered*)—but that alone does not (in anyone's book) make them count as artifacts, and their subjection to these norms need not be essential to their membership in the kind *cow* or *stone*. Members of public extant artifact kinds are distinguished from other things ready-to-hand in that not only are they subject to public norms but that it is an *intended* feature, essential to membership in the kind, that they be subject to those norms.

What I would like to draw out of these diverse sources is a single general idea: that the definitive intended properties for membership in public artifact kinds typically include not merely functional or structural features, but also being intended to be subject to certain norms, where this is understood as the object being recognizable (by an intended audience) as to be treated, used, regarded, etc., in certain ways. Makers intend their creations to be recognized by an appropriate audience so that that audience may treat them properly, subjecting them to the relevant norms

[7] Heidegger of course is no friend of talk about individual intentions, but we may nonetheless find much of use in his way of understanding the defining features of the objects we live and work with.

[8] "So in the environment, certain entities become accessible which are always ready-to-hand, but which, in themselves, do not need to be produced" (Heidegger 1962, p. 100).

regarding how the object created is *to be* treated or regarded, how, in what contexts, and by whom it is *to be* used, considered, behaved regarding, etc. To intend to make a work of art, a cathedral, a cheese sauce, or a top hat, is (inter alia) to intend to make something that is *to be recognized as subject to certain norms of use, treatment, regard,* etc., by an appropriate (intended) audience.

These norms may include not merely how the artifact itself is to be treated, but also how its bearer, user, or other things to which it is related are to be treated: Thus, e.g., uniforms impose not only norms about how *they* are to be used (how they are to be worn, on what part of the body, by whom, in what circumstances), but also norms of behavior for the wearer (consider the soldier's uniform) and for those who interact with the wearer (consider the police officer's uniform). Price tags come with norms of treatment for themselves and for anything they are (properly) affixed to. Buildings, as mentioned above, come with norms for how those who enter them are to behave, and so on.

It is important to note that these must be understood as genuine *norms*, not just regularities involving what people (descriptively, happen to) *do.* Someone who *misuses* a mechanic's tool or behaves *improperly* in a church (by acting in ways *proper to* a gymnasium) is subject to *correction* or *rebuke*, signs that *norms* (not merely regularities) are at issue. Even those who fail to use their eating utensils "properly" (relative to the context) are subject to scorn and correction, regardless of how successful they are at using them to efficiently shovel food. So we cannot see this correction merely as having the form of a conditional suggestion: If you want to eat more efficiently, use your chopsticks (or fork) this way. Instead, the correction is designed to show the user the *proper* way to eat with chopsticks (or a fork).

We can perhaps uncover a deeper norm regarding treatment of all artifacts: that artifacts are (prima facie) *to be treated as* their creator *intends them to be treated.* (This of course is not to say that those norms cannot be overridden by other norms of politeness or morality, e.g., if the object is a noisemaker or a weapon). For this reason, the norms an artifact *actually is* subject to and those it is *intended to be* subject to generally coincide.

They can come apart, however. Where they do, we tend to classify artifacts by way of the norms they are *intended to be* subject to in their *intended context,* not those they are actually subjected to in their actual context. Cases of exaptation occur when an artifact is successfully created with the intention that it be recognizable as subject to one set of norms (by an intended audience), but is in fact treated in accord with other norms. Suppose, for example, that we are in a state of isolation from the Chinese community—except for trade—and that (expecting we eat as they do) the Chinese export to us a large number of chopsticks. Having no idea of their practices, but having practices of wearing and tying up long hair, people in our society purchase them for use as hair sticks (and perhaps thinking that they are hair sticks). Intuitively, it seems that these are still chopsticks: They are the products of intentions that they be recognizable as chopsticks and so subject to chopstick-related norms—e.g., that they are to be used in eating. And they *are* so recognizable by the *intended* audience, though they have been taken out of context. The only problem is that, given their uninformed actual audience, they fail to be actually recognized as subject to the intended norms and are treated in accord with other norms instead (those governing hair accessories). Here, it is clearly the norms the objects are intended to be subject to, rather than those in accord with which they are actually treated, which play the lead role in classification.[9]

In any case, to say that members of the relevant artifact kinds must be intended to be *subject to* certain norms (in the sense of being recognizable as *to be* treated in certain ways rather than others) of course does not entail that members of the relevant artifact kind *will be* treated in accord with those norms—even in their intended or "home" context. The presence of norms does not prevent the intentional violation of norms. But you cannot willfully violate norms without recognizing them. To desecrate a flag requires recognizing it *as* to be treated in certain respectful ways and intentionally violating those norms of treatment. And school children who fling rice pudding on the ceiling as a protest about their dessert show that they have recognized it as *to be eaten* even while they refuse to follow that norm. (It is the norm: that they *are supposed to eat* that stuff, rather than the pudding itself, that they resent. Had the same material been presented as part of a science project, it likely would have aroused no such resentment and misbehavior). So the fact that norms governing their use may be flouted does not undermine, but instead presupposes that public artifacts are subject to norms.

4.4 Dependence on Public Norms

Artifacts are standardly treated as mind-dependent entities, since for an artifact to be created, there must be fairly structured intentional states, involving an individual intending to make a thing of a certain sort, with certain intended properties—and also, of course, being relatively successful at executing those intentions.

It is often supposed that artifacts are the expressions of *individual* intentions and actions, and as such, depend merely on the intentions of those individuals who make them. So, for example, I have argued elsewhere that artifacts differ from properly social and institutional objects in *not* depending on collective intentionality: "Unlike social and institutional objects, the existence of artifacts doesn't seem to presuppose any *collective* intentions of any kind – it makes perfect sense to suppose that a solitary human could create a knife, though not a government or money" (2007, p. 52).

[9] Are they (also) hair sticks? Here, I think (if properly informed about the origins and home use of such things), we'd naturally dither – we might say: "They were meant to be chopsticks, but we use them as hair sticks" – or "they're hair sticks to us." We can of course also allow that in our context, we can engage in a kind of minimal making of a new kind of artifact (hair sticks) exapting the prior ones as – in this context – there are no conflicting norms of use for these things to interfere with others recognizing my imposed norms of use on these things (when I use them to hold up my hair). But to the extent that we think of them as hair sticks, I think, we are thinking of Westerners who adopt them for this purpose as engaging in a kind of minimal making of a new artifactual type, intending them (placed in the proper context: the hair care aisle) to be recognizable as subject to new norms.

Even if we understand public artifacts as I suggested above, that might still seem in principle to allow us to hold that these artifacts depend only on the intentions of their makers, even if these are nested intentions regarding the intentions of others (intending that others see or use the product in a certain way).

I have begun to think, however, that this is not so—that there is an important and revealing sense in which members of public artifact kinds *do* depend on intentional states beyond those of their makers. The need for individual intentions alone does not seem to fully capture what it is to be a member of one of our standard, extant artifact kinds: what it is to be a table, a teapot, or a salad fork. We might say that a member of the extant artifactual kind *salad fork* is not just something this guy made so he could eat his salad with it, but rather something *successfully* created with the intention that it be recognizable as something *to eat salad with*. What is it to intend something to be recognizable as *to eat salad with?* Something along these lines seems right: that it be intended to be recognizable as subject to certain *norms of use*—that is *to be used* for a certain purpose (eating salad), in a certain context (where there is full "dinner service," including larger forks), in a certain way (by holding the handle in one hand, turning it in a certain way, stabbing (not shoveling) the food…), and by certain individuals (grown people: the size here not being indicative that it is for children). For such an intention to be even moderately successful, then, there must *actually be* such established public norms in place (so that the maker can intend that this be subject to those norms of treatment).

But what does it take for there to be public norms like these? This, of course, is a major question in itself and cannot be resolved here. Nonetheless, it is fairly clear that public norms of use cannot be established simply by the individual intentions of an artifact's maker. Instead, there must be widespread intentional states within the relevant society, of people who accept, recognize, or consider things like these as things that are to be treated, used, or regarded in certain characteristic ways (and perhaps who do so in conditions of common knowledge).

This is good reason for thinking that members of public artifactual kinds depend on mental states beyond those of their maker; but does it also show that they depend on *collective* (not just individual) intentionality? While that seems plausible, given the ongoing rich debates about how to understand collective intentionality, it is not a question that can be answered definitively without first answering difficult questions about how to understand collective intentionality.[10] These difficulties are compounded by the fact that the main target of analyses of "collective intentionality" has been to understand what it is for a group (e.g., a corporation, a team) to share an intention to act together, or to share beliefs or responsibility—not to understand the kind of diffuse societal recognitional support of norms that is at issue here.

Nonetheless, without settling debates about how to understand collective intentionality, we can make the core point at issue here: that the existence of members of

[10] For example, debates concern whether we should understand collective intentionality in terms of we-form intentions in individual minds (Searle 1995), in terms of individual states related in the right sorts of way (Bratman 1999), in terms of states attributable to plural subjects (Gilbert 1996), etc. For a helpful summary of the debate, see Tollefsen (2004).

our extant public artifactual kinds (though perhaps not of private tools) relies on the existence of public norms—and thus also on *whatever* sorts of intentionality are needed for such public norms to exist (leaving this to future research). More specifically, we might say, there are constitutive norms for being a knife or a table or a teapot, and the existence of objects of these kinds (in the full-blown, meaningful public sense—not just the "tool" sense) relies on the presence of such norms.

Two kinds of apparent counter-evidence might be raised against this claim. First, it does (as I earlier (2007) alleged) seem that a person on a deserted island may in a sense create artifacts—things with certain intended properties for her own use, and that we might investigate these things as artifacts, asking what they were supposed to be, what their intended use was, and so on. Second, it also seems that an inventor *within* a society may create a prototype of a new sort of artifact, characterized by various intended features, and that its status as an artifact is guaranteed by its being the product of intentions to create an object with those very features.

But even these objects may, in the normal or core cases, depend on the presence of public norms. Most inventions are not completely ex nihilo, but rather based on prior and broader types of artifactual kinds, so the Wright Brothers may well have intended their creation to be subject to at least some of the norms for treatment of transportation devices (rather than, say, those for treatment of religious artifacts or works of art), and so being subject to certain norms of treatment may come into play as an intended feature even for novel kinds of artifact.

In the most natural way of imagining the case of the desert islander, we must imagine that she arrives on the island beyond the age of infancy (or else she would have had no chance at surviving), and so as already to some degree enculturated—of an age to have known what a house, knife, plate is, how they are to be used, etc. If so, it may be that her intention to create a house *is* an intention that it be the kind of thing in principle recognizable as subject to those norms (even if no one else is around to recognize it or follow those norms). If we think of the desert islander as capable of making a member of the *public* artifact kind *house* or *teapot*, she must have come from a culture in which there were the relevant kinds of house-regarding and teapot-regarding norms, so that she can intend her products to be recognizable as the sort of thing subject to those norms. And thus her ability to successfully make a house or a teapot in this full-blown sense still relies on these public norms. On the other hand, if she arrived as an infant, was raised by wolves, and miraculously survived and made something to sleep in or make water flavored with leaves in, her products may be independent of public norms, but these will not be public artifacts in our sense, but rather private tools (which might still be counted as "artifacts" in a broad sense).

4.5 Virtues of the Analysis

I have argued that what is essential to the existence of an artifact is that it be the intended product of human activity, and that for an artifact to come into existence, it must be intentionally created and successfully endowed with certain intended features—intended features that may, but need not, include an intended function. Moreover, I have argued that artifacts are classified into artifactual kinds by way of their possessing intended features criterially associated with that kind—and that these classification-relevant features again may, but need not, include intended function. I have also emphasized that such intended features may and often do include intended recognizability as to be treated, used, regarded, etc., in certain ways (by a certain intended audience, in a certain context). For such normative features to be successfully imposed in turn requires that there really be such public norms of treatment.

The idea that members of public artifactual kinds typically have intended subjection to norms among their essential classificatory features has many virtues. First, it seems to unite the treatment of functional, religious, and art objects—a problem that has often arisen for other theories of artifacts. (Indeed, getting works of art to fit seamlessly into a theory of artifacts has been an enduring challenge.) If we take the relevant intended features unifying artifacts into artifactual kinds to be functional, we risk leaving out works of art; if we make them structural or perceptible, we seem to get a poor view of many functional artifact kinds (which can vary widely in structure) and also an implausible view of art kinds (which seem not to be definable in that way). But allowing complete latitude over the relevant intended features and acknowledging the core role that intended receptive features may play promises to unite these all in a single account.

Indeed, the intention that the object be subject to certain norms seems in many ways most central and basic (though it is easy to overlook). The intended *functional* features are also tied to norms about what the object is *to do* and how (and by whom) it is *to be* used to achieve this function. Moreover, as suggested above, the intended *structural* features of artifacts often serve in part to make the type of object *recognizable* by the intended audience, so that it can call forth the appropriate norms. In other cases, intended structural features may serve more directly to call forth norms. Many structural features of items of clothing, for example, have nothing to do with the function of covering the body or retaining warmth or even with making it recognizable as a shirt or jacket, but rather are aimed at calling forth certain norms of behavior towards—and from—the wearer. Consider the relevantly different norms called forth by the structural and perceptible differences say, between ratty gym wear and a designer suit. (Schools have long noted the force of these norms in affecting the behavior of children and used this as one argument in favor of having children wear nice uniforms or "dress up" for a dance). Uniforms of all sorts (for employees, police officers, etc.) even more obviously serve the role of calling forth appropriate norms of behavior from wearers and observers.

We also gain an important benefit by considering members of public, extant artifactual kinds as dependent on public norms: It helps us address the exaptation problem for artifacts—a problem that has particularly arisen for intentionalist conceptions of artifacts. Here is the problem: Can you make an artifact "minimally"—without changing anything about it? Certainly sometimes it seems like you can, e.g., make a river stone a paperweight just by placing it in the proper context and intending it to be your paperweight. But then what do we do about exaptation;

when I adopt, say, a teapot as a paperweight—have I then made a new sort of artifact (a paperweight, co-located with or replacing the teapot)? It does not seem so. But to deny that seems arbitrary, if we allowed creation in the case of the river stone.

One way to address this problem is to allow that in both cases I may have (minimally) made a new private tool—but only in the first case have I made something of the public type "paperweight." Why? In the first case, by placing something of that size and shape in that context, I may also legitimately intend others to recognize it as a paperweight to be subjected to the relevant norms, and my intention may plausibly be successful. The rock in itself is innocent of artifactual norms; by putting it in the proper context, I can successfully impose some new norms—in accord with public practices, which do involve treating things of that sort of size, shape, and material, found in an office context, as paperweights (rather than as fire stones, food mashers, weapons…). If you pick up my (now obvious) paperweight and use it to mash potatoes instead, I might justly complain—and (provided you have recognized it as a paperweight) you will probably only do so with a sense of violation (whether in glee, anger, or indifference).

In the second case, if I intend the teapot to be recognizable as subject to paperweight-regarding norms without changing anything about it, I will likely be unsuccessful: Given the object's structural features and context, others will see it as to be used for making tea and subject it instead to teapot-regarding norms. If I naively intend it to be recognizable as subject to paperweight norms instead, without in any way making this clear (through modification, signage, etc.), I will fail; other norms are in place that prevent the intended audience from recognizing those I hoped to impose (unless important, crucial changes are made). (It is something like trying to use "cat" to mean "dog.") Our receptive intentions, like our functional intentions or structural intentions, may fail—and if they fail badly enough, and if those intended features are core criterial features for membership in the kind, our attempt to make an artifact of that type may fail.

Here is another puzzle: Many have the intuition that one can make something an artifact, or even a work of art, just by way of selection and display—and this seems in accord with some practices in the art world regarding found art and the like. So, for example, George Dickie holds that one may "confer" artifactuality on an object, like a piece of driftwood, merely by hanging it on the wall. "Natural objects which become works of art… are artifactualized without the use of tools – the artifactuality is conferred on the object rather than worked on it" (1971, p. 106). Yet, even if that seems right, as Jeffrey Wieand has pointed out, it seems that one cannot confer artifactual or art status on the same piece of driftwood if it is simply lying on the beach: "Someone who tried to confer status on a piece of driftwood lying on the beach would be trying to confer status on the wrong sort of thing" (Weiand 1980, p. 386).

If this distinction seems right, we can now see why: On the beach, the right norms are not in place for a piece of driftwood to be recognized as to be subjected to art-regarding norms. The intention that it be so recognized will likely fail (unless the circumstances are special: the person attempting it is a well-known artist; the beach is Miami Beach during Art Basel week…). But once the driftwood is hung on the wall, especially in a museum (with proper signage etc.), it can be recognized as *to be treated* as art is to be treated, and so intentions that it be subject to those art-regarding norms can, in this circumstance, be successful.

4.6 The Interest of Artifacts

A further virtue of understanding public artifact kinds as dependent on public norms of treatment for members of the kind is that it can give an adequate account of why artifacts are of interest to history, archeology, and other social sciences. Taken solely as use-objects or tools, they might indeed be interesting to engineers—as objects intended to serve (and perhaps serving) a particular use. But that is not the main interest social scientists take in human artifacts. Dipert writes "Artifacts are the 'residue' of intentional activity" (1993, p. 15); given the above considerations, we can broaden this to say that artifacts (at least of public, extant kinds) are the residue of human intentions *and normative practices*. They are not merely natural objects like any others, nor merely objects that may have a certain (intended) use. To see public artifacts *as* artifacts is to see a way of life—to take interest in public artifacts as such is to take interest in what they were made for, what norms and practices governed things like that, in the context of the broader practices and setting of a society. It is for this reason that they are of particular interest to historians and archeologists and are earnestly preserved in museums: as capable of providing particular insight into a way of life that may be distant in time or place. Anthropologists and archeologists are not just interested in the physical object, nor in a physical object some individual intended to have certain features and perhaps to use in a certain way, but rather in what *these objects were seen as, what they were for, what norms of use they came with, who they were to be used by, and how they were to be employed as part of a way of life.* That is also why artifacts can be creepy, when part of a civilization is lost or destroyed and we see, e.g., the remnants of Pompeii under the volcanic dust and thereby discover not just physical things of interesting shapes, but a lost way of life.

Understanding artifacts in this way also provides the potential for a better account of the full significance of artifacts and their roles in our lives—which go far beyond the roles of mere tools that we hope will help us get things done. As Stephen Laurence and Eric Margolis put it:

> …human artifacts aren't purely utilitarian objects. They also have enormous cultural value. There is a big difference between driving a Volkswagen Beetle and a Hummer, or between wearing the latest Armani suit and an old pair of ripped, baggy jeans. The artifacts we surround ourselves with speak volumes about what is important to us, what groups we identify with, and who we are as individuals. (Margolis and Laurence 2007, p. ix)

Acknowledging the role that public norms play in the very existence of public artifact kinds can help make these kinds of distinction far better than mere functional or structural accounts alone could. The artifacts we surround ourselves with impose certain norms of behavior on us (which of course we can always willingly choose to violate) and on those who interact with us—who see our clothes, park our cars, and visit our homes.

In short, on this view, public artifacts turn out to be dependent on public *norms* regarding how we are to treat them, to behave regarding them, and so on; they build in reasons for acting in some ways rather than others. The role of artifacts in our lives is as much about settling us into a web of norms as it is about achieving more basic practical goals like warmth and transit.

4.7 Artifacts, Art, and Language

Certain works of art also exploit and make evident the importance of public norms on membership in artifact kinds. So, for example, Duchamp's "Fountain," given its exact similarity to a urinal, invokes certain use practices that are then also forbidden by its improper context: its placement in an art gallery. Meret Oppenheim's "Fur-lined teacup" similarly invokes by its shape certain norms of use (for sipping warm beverages) that clash with other structural features (its being lined with fur), and Claes Oldenburg's enormous clothespin, safety pin, and trowel invoke by their shape norms of use that clash with the usability of the objects by making them of entirely the wrong scale for normal (intended) users. All of these works derive at least part of their interest from the way in which they exploit a clash between our standard criteria for membership in artifactual kinds: By their obvious and apparently intended structural features, these objects invoke recognition that they are to be treated as members of the kind, but by their placement, form, or size, they also intentionally prohibit following such norms and impose others instead (the norms of behavior regarding art).

Perhaps the most interesting is Felix Gonzales-Torres' "Untitled" (Portrait of Ross in L.A.): a "portrait" composed of a heap of candies, with a sign inviting visitors to eat the candies, and instructions to curators to replace them. This work invokes the norms of use of candies as to be eaten, norms which are contravened by the norms of art museums[11] (do not touch, still less eat, the works), and then again contradicts those norms by adding a sign encouraging viewers to go ahead and eat the constituent candies. In this case, we have a double play with the relevant norms.

In sum, acknowledging the role of intended normative and recognitional properties in constituting our standard, public artifactual kinds brings with it a number of advantages. It enables us to provide a better account of artifactual classification and to better handle cases of minimal making and exaptation. It also provides a way of understanding the ways we experience artifacts (as exerting a kind of normative pull on us), and the reasons we take social-scientific interest in them *as* artifacts. It can even help to understand the interest and power of certain works of art.

[11] Rather (by placement and arrangement), the work invokes art-regarding norms and retains some (e.g., that it is to-be-interpreted, to-be-contemplated), while it rejects others (by adding a sign suggesting it as to-be-eaten).

A final point of interest arises from acknowledging the dependence of members of extant public artifact types on public norms is the commonalities that can thus be seen to arise between artifacts and language—at least if we take a roughly Wittgensteinian approach to language. Artifacts (like words) are inherently *meaningful* (where meaning is considered as rules/norms of use); artifacts (like language) also must be understood *holistically* (like words, they have significance and are what they are only as part of a total context—whether of other artifacts or of other words); and artifacts, like words, are fully understandable only against the background of a way of life (range of norm-involving practices). These commonalities which come to the fore given the above understanding of artifacts may open up the way to seeing commonalities between language and artifacts—both with a view to seeing artifacts as a meaningful part of culture (rather than seeing them as mere things with certain physical-functional capacities) and to seeing language as just one particularly interesting cultural artifact.

Acknowledgments Many thanks to Simon Evnine, Beth Preston, and the editors for very helpful comments on an earlier draft of this chapter.

References

Baker, L. (2008). The shrinking difference between artifacts and natural objects. In P. Boltuc (Ed.) *Newsletter on Philosophy and Computers. American Philosophical Association Newsletters, 07*(2), 2–5.

Bloom, P. (1996). Intention, history and artifact concepts. *Cognition, 60*, 1–29.

Bratman, M. (1999). *Faces of intention*. Cambridge, MA: Cambridge University Press.

Dickie, G. (1971). *Aesthetics: An introduction*. Indianapolis: Bobbs-Merrill.

Dipert, R. (1993). *Artifacts, art works and agency*. Philadelphia: Temple University Press.

Gilbert, M. (1996). *Living together: Rationality, sociality, and obligation*. Lanham: Rowman & Littlefield.

Heidegger, M. (1962). *Being and time* (J. Macquarrie & E. Robinson, Trans.). San Francisco: Harper.

Hilpinen, R. (1992). On artifacts and works of art. *Theoria, 58*, 58–82.

Hilpinen, R. (1993). Authors and artifacts. *Proceedings of the Aristotelian Society, 93*, 155–178.

Hilpinen, R. (2008). Artifact. In E. N. Zalta (Ed.) *The Stanford encyclopedia of philosophy*. Stanford: Stanford University. http://plato.stanford.edu/archives/fall2008/entries/artifact/

Husserl, E. (1989). *Ideas pertaining to a pure phenomenology and to a phenomenological philosophy. Second Book* (R. Rojcewicz & A. Schuwer, Trans.). Dordrecht: Kluwer.

Ingarden, R. (1989). *The ontology of the work of art*. Athens: Ohio University Press.

Kornblith, H. (1980). Referring to artifacts. *Philosophical Review, LXXXIX*(1), 109–114.

Levinson, J. (2007). Artworks as artifacts. In *Creations of the mind: Theories of artifacts and their representation* (pp. 74–82). Oxford: Oxford University Press.

Margolis, E., & Laurence, S. (2007). Introduction. In *Creations of the mind: Theories of artifacts and their representation*. Oxford: Oxford University Press.

Millikan, R. (1984). *Language, thought and other biological categories: New foundations for realism*. Cambridge, MA: MIT Press.

Preston, B. (2009). Philosophical theories of artifact function. In A. Meijers (Ed.), *Philosophy of technology and engineering sciences* (Vol. 9, pp. 213–233). Amsterdam: Elsevier.

Searle, J. (1995). *The construction of social reality*. New York: The Free Press.

Thomasson, A. L. (2003). Realism and human kinds. *Philosophy and Phenomenological Research, 67*(3), 580–699.

Thomasson, A. L. (2007). Artifacts and human concepts. In E. Margolis & S. Laurence (Eds.), *Creations of the mind: Theories of artifacts and their representation* (pp. 52–73). Oxford: Oxford University Press.

Tollefsen, D. (2004). Collective intentionality. *Internet Encyclopedia of Philosophy.* http://www.iep.utm.edu/coll-int/

Weiand, J. (1980). Defining art and artifacts. *Philosophical Studies, 38*, 385–389.

Chapter 5
Artefact Kinds, Ontological Criteria and Forms of Mind-Dependence

Maarten Franssen and Peter Kroes

Abstract In this chapter we discuss criteria for ontologically crediting or discrediting certain kinds of things we refer to in everyday life and artefact kinds in particular. Generally used criteria for settling whether things 'really exist' are mind-independence and determinateness, and on these criteria artefacts are said to flounder. We show that another criterion, which we term the phase-substance criterion, is also of relevance for delineating what are the real kinds of things in the world and what are the merely nominal kinds. We use these criteria to argue that artefact kinds can be defended as real in the form of intentional-historical subkinds of structural kinds. We show that the relation between these structural kinds and their intentional-historical subkinds is mirrored by a similar relation for natural kinds in biology and that similar forms of a division of explanatory labour are at work in both cases.

Keywords Artefact • Natural kinds • Nominal kinds • Ontology • Real kinds • Species

5.1 Introduction: The Ontological Status of Artefact Kinds

If you use a car to drive home from work, and use a key to unlock the car door, you take these things to be real, as real as the tree you wish to avoid when parking your car. The anxiety you feel when you face a car approaching you at high speed is not different from the anxiety you feel when a growling dog is chasing you. So how can it be that some philosophers call the reality of any of these things into question?

M. Franssen (✉) • P. Kroes
Section of Philosophy, Delft University of Technology,
Jaffalaan 5, 2628 BX Delft, The Netherlands
e-mail: m.p.m@tudelft.nl; p.a.kroes@tudelft.nl

M. Franssen et al. (eds.), *Artefact Kinds: Ontology and the Human-Made World*, Synthese Library 365, DOI 10.1007/978-3-319-00801-1_5,

And even more strange, how can it be that some philosophers call into question the reality of artefactual things like cars and keys but not the reality of natural things like dogs and trees?

To be sure, hardly any philosopher nowadays puts into question the reality or existence of something material being there and being able to causally affect the equal materiality of our bodies. What is put into question is the existence of entities of specific *kinds*, their 'being-givenness' as wholes, at various levels of generality, which includes the reality of artefacts qua artefacts, in distinction to the existence of particular configurations of matter at the locations where everyday usage takes these artefacts to be present. In this chapter our aim is, first, to make explicit criteria for discriminating, among all the things that our everyday language seems to acknowledge, between those that are 'really' part of the 'furniture of the world' and those that are 'unreal' or second order in some sense or other, and, second, to discuss to what extent these criteria succeed in doing so. We start with two preliminary sections. In Sect. 5.2 we argue that it is the *reality* of kinds, not the *existence* of kinds themselves, that is at issue. With respect to this reality issue, we argue in Sect. 5.3 that we lack a definite criterion that discriminates straightforwardly between ontologically serious and ontologically less serious kinds of things. As a consequence, we must accept either a much sparser or a much richer ontology than common sense has it. In Sect. 5.4 we shift the emphasis to artefact kinds and show that the contrast between artefact kinds and natural kinds of things is less clear-cut than is often assumed. In Sect. 5.5 we argue that artefact kinds cannot be introduced directly as functional kinds, irrespective of whether an artefact's function is conceived in terms of designer intentions or conceived in terms of physical capacities. Artefact kinds can be introduced, however, in the form of structural kinds, similar to natural kinds, which are further specified by adding, in a way presented in Sect. 5.6, an intentional-historical dimension, the artefact's origin in designer's and manufacturer's intentions or creative practices. In Sect. 5.7, we argue that artefact kinds so conceived show a similarity to natural kinds, by drawing a parallel to the status of animal kinds and species in biology. In our closing Sect. 5.8, we discuss what has been achieved and draw some relativizing conclusions.[1]

5.2 Real Things, Real Kinds and the Existence of Kinds

Our first step is to investigate what is at stake in existence claims with respect to certain kinds of things. The standard cases where we say that 'something' does not exist or is imaginary are cases where the sort of thing discredited is similar to, of the same type as, other sorts of things that are accepted as existing in an ontology that accepts the existence of things of various kinds. UFOs, zombies and the monster of

[1] In the following we are primarily interested in the ontological status of technical artefact kinds and not of artefact kinds in general; whenever we refer to artefacts or artefact kinds, this should be read as technical artefact or technical artefact kinds.

Loch Ness are supposed to be (by anyone who upholds their possible existence) material objects similar to people, animals or aeroplanes. Their non-existence means that there is no aggregate of matter in the world that at any moment corresponds to, or makes up, any of these things. When the existence of artefacts, or of ordinary objects in general, is put into question, what is at issue is not their existence in the way that the existence of the monster of Loch Ness or of zombies is questionable. When we use a screw driver, there is something material that we physically interact with by the way we hold and move our hand; that much cannot be put in doubt by anyone who accepts the existence of a material world.

In our discussion, we make no distinction with respect to individual things between their reality and their existence. With respect to the reality and existence of *kinds* of things, however, we do make a distinction. Speaking of the reality of a kind is to express that things as instances of that kind exist, and when the existence of things of a particular kind is accepted, that kind is taken to be real. Thus, the *reality of a kind* implies the reality/existence of its instances and the reality/existence of instances of a kind implies the reality of its kind. These are two ways to express the same thing. Speaking of the *existence of kinds*, in contrast, is to express that kinds are themselves some sort of thing, which may or may not fall under higher-order kinds. To accept kinds as existing amounts to accepting the existence of abstract things, things not located in space and time. The position that universals exist as individuals is traditionally referred to as realism, whereas nominalism is the position that only their instances exist, but not the kinds themselves. Although it is not accidental that the word 'real' is used both for branding the position of realism with respect to the existence of kinds and for branding 'real kinds' as kinds whose members exist, and similarly for 'nominal', we will not assume that accepting a distinction between real and nominal kinds, in the sense defined above, implies anything about the acceptance of either realism or nominalism in the debate about the existence of universals. With regard to the problems discussed in this chapter, we consider it irrelevant which answer is given to the question of the existence of universals. In this chapter, the *existence of individual material things as instances of certain kinds* is the issue (which issue is identical to the issue of the reality of those kinds).

Denying the existence of artefacts can occur in two forms. (1) There are those who deny the existence of artefacts as included in their denial of the existence of the totality of ordinary objects; what is denied, more specifically, is that any macroscopic material object can reasonably be ontologically classified as a particular kind of thing. (2) There are those who accept the existence of many ordinary kinds of things, for example animals, but deny the existence of (but perhaps not exclusively of) artefacts. What is denied is that something material can reasonably be ontologically classified as an artefactual kind of thing.

The former, more radical position, defended by, e.g., Unger, Merrick and, to some extent, van Inwagen, has recently been attacked by Thomasson (2007a). The proponents of this position typically introduce kinds of things only in terms of the theoretical notions of physics: what exists in the case of ordinary things are exclusively collections or aggregates of atoms and molecules of particular kinds, but not things that are instances of macroscopic kinds. Thomasson argues that to accept the existence of an aggregate of molecules in the form of a baseball is to accept the existence of a baseball at the location where the aggregate is. For a baseball to exist, certain existence and identity conditions must be fulfilled, and these are in fact fulfilled if an aggregate in the form of a baseball is present. That is simply what we mean when we say that there is a baseball at a particular location.

Thomasson develops this into a very liberal ontology. A particular thing of a particular kind, referred to by a particular sortal, exists if something material satisfies the application and coapplication conditions of the sortal, to which correspond the existence and identity conditions of things of that kind. Thomasson specifically argues that this position does not amount to 'thinking things into existence'. We are free to propose existence and identity conditions, but it is up to 'the world as it is' whether there are things that fulfil them. A consequence of this position is that we must accept the reality of a great many kinds of things that we do not normally consider as 'being around'. An example (suggested by van Inwagen) are gollyswoggles, which are "lumps of clay with a particular very complicated shape". Their apparent uncanniness merely originates from the fact that these things play no role in our lives or in our scientific theories. We can, however, make room for them in our language, learn to identify them, count them, and so forth, in the same way that we learn this for commonly accepted things like tigers and tables (2007a, pp. 172–173, 183–185).

The latter of the two positions sketched above, in contrast, has been addressed equally recently by Baker (2007). She distinguishes ordinary things like bicycles and desks, which are instances of real kinds, from, in her view, pseudo-objects like 'penny-in-a-pocket'. Bicycles and desks exist, but not so pennies-in-a-pocket. She does not deny that there are pennies – as there are desks and bicycles – which can temporarily be located inside a pocket, and therefore have the property of being in a pocket. But when a penny is put into a pocket, the world does not become richer, as she puts it, whereas it does become richer when a bicycle is assembled. This view immediately raises the question on what grounds things like bicycles can justifiably be separated as real from – in Baker's view – pseudo-things like pennies-in-a-pocket. Can we come up with an ontological criterion that is generous enough to give us common artefacts as serious things but not so generous as to allow the existence of gratuitous kinds of things like pennies-in-a-pocket? Confronting this with Thomasson's view, it seems there is nothing that prevents us from stating existence and identity conditions for things of the kind 'penny-in-a-pocket' and from checking that these are fulfilled by pennies that are inside pockets. And indeed, we have no difficulty counting the number of pennies-in-a-pocket there are in a room.

What Baker wishes to deny by denying that there are such things as pennies-in-a-pocket is, in ontological terms, that pennies-in-a-pocket are members of a kind of thing in itself, different from the kind 'penny' but of equivalent ontological status. In Baker's view, there is something ontologically frivolous about pennies-in-a-pocket; the kind 'penny' is an ingredient of the world, but not the kind 'penny-in-a-pocket'. The general issue here is a distinction between *real kinds* and *nominal kinds*: only the things corresponding to real kinds 'really' exist, as we tend to express it, whereas nominal kinds correspond to divisions that 'exist' only in our minds but lack any correspondence to what is 'out there'.

Given our distinction between things that exist as instances of a particular real kind and things that merely fall under a nominal kind but exist as things of other kinds, a crucial question becomes how we can distinguish between real and nominal kinds. By what criteria can we accept certain things as existing qua thing of a particular kind and can we reject other things as not 'really' a thing of its kind or reject that kind as not 'really' a kind? And what price do we have to pay for these criteria? Baker herself fails to present any argument that can ground her inclusion of bicycles and desks and the exclusion of pennies-in-a-pocket from what exists. In general she says that a member of a kind F comes into being when 'F-favourable circumstances' obtain, but gives no clue to the sort of circumstances that are favourable to the generation of a member of a kind in general or an argument why there are bicycle-favourable circumstances but no 'penny-in-a-pocket'-favourable circumstances.

5.3 Reality Criteria and the Problem of Proliferation

Contrasting Thomasson's and Baker's positions has introduced us to the important question what restrictions we can or must accept on what exists as an instance of a real kind. One of the most often referred to intuitions is that existing as an instance of a real kind cannot depend on someone's thoughts, and therefore we should only accept as real things that what does not depend on anyone's thought. This is often interpreted as implying that artefact kinds are nominal kinds and that artefacts do not exist, since they are commonly seen as depending on human intentions, in particular the intentions of their designers or makers and of their users. This consideration, however, seems to be orthogonal to the considerations of Thomasson and Baker. Both straightforwardly accept artefact kinds as real, while acknowledging that being an instance of an artefact kind crucially depends on human intentions.

For Thomasson, although it is not clear what restrictions she accepts, if any, to existence and identity conditions for things as instances of a particular kind, it is clear that a dependence on intentional states is not among them. She characterizes an artefact of kind K as the product of some person's largely successful intention to realize a substantive concept, had by this person, of the nature of Ks that matches that of prior makers of Ks (if any), by imposing K-relevant features on the object in question (2003, p. 599; see also her 2007b). Neither can a dependence on intentional states ground Baker's drawing of the dividing line between real and nominal kinds. Similar to Thomasson, Baker holds that artefact kinds depend essentially on mental states, and she explicitly rejects mere dependence on mental states as a criterion for the ontological disqualification of kinds, or at least she does so in the case of artefacts (2007, pp. 63–64). And indeed, the exclusion of 'penny-in-a-pocket' as a real kind cannot have anything to do with intentionality. The kind 'pebble-on-a-rock', for whose instances the existence and identity conditions do not depend at all on the existence of people, would, in the eyes of anyone who rejects 'penny-in-a-pocket' as a real kind, be just as ontologically suspect.

We could decide that restrictions on what are real kinds can be grounded in just that, the intuition that there must be restrictions on what can count as a real kind, because if we do not place restrictions somewhere, we will be confronted with a counter-intuitive proliferation of real kinds of things. Once we accept 'penny-in-a-pocket' as a kind, it becomes difficult to deny kindhood to 'penny-in-a-trouser-pocket', 'pair-of-pennies-in-a-pocket', 'penny-in-a-pocket-on-a-Tuesday' and so forth and so on. But to intuit a need for criteria on ontologically respectable kindhood is not to have such criteria or to know how to apply them. Thomasson accepts that there may well be instances of millions of kinds of objects on her desk. She suggests, however, that the conflict with common sense is only apparent, because the concept of 'thing' under which there are millions of kinds of things on her desk is a "covering concept", which is different from the common-sense concept that understands things as "cohesive, enduring, medium-sized separate physical entities". This distinction seems to us not to be relevant here; however, a gollyswoggle undeniably is a cohesive, enduring, medium-sized separate physical thing, so even of things in the common-sense interpretation there may well be instances of a great many kinds of things on her desk. If she accepts this, then apparently Thomasson finds the proliferation of things in existence a price worth paying for saving the kinds of things she wishes to save – ordinary objects, including artefacts. She concedes that the totality of existing (in our terminology 'real') kinds of things is given by the totality of possible concepts that happen to be realized (2007a, pp. 124–125) and that it is highly indeterminate. This is an ontology that many would consider too rich.

Let us, therefore, see whether there is a criterion that can discriminate between the kind 'penny' and the kind 'penny-in-a-pocket'. The latter seems close enough to 'red table', which is a sortal all right – we can count the number of red tables in a room – but red tables seem a paradigm case of things we would not want to accept as instances of a real kind: if we did, we would have to accept the painting of a red table green as the going out of existence of a particular red table. We prefer to have 'red table' indicate a form under which instances of the real kind 'table' can occur. Wiggins (1980, p. 24) calls such sortals 'phase sortals', since they indicate a phase in the career of individuals of some sort, against 'substance sortals', like 'tiger', which name real kinds. Hirsch (1982) has proposed a criterion for discriminating between substance sortals and phase sortals. The starting point is the observation that very little falls under precisely one sortal: a tiger is also a mammal or an animal, a penny is also a coin, and a penny-in-a-pocket is also a penny. Between these sortals there are various dependence relations, one of which Hirsch singles out by defining one sortal to be *subordinate* to another sortal if necessarily, if we have an instance of the former, we also have an instance of the latter. Thus, 'tiger' is subordinate to 'animal', 'red table' to 'table', 'penny-in-a-pocket' to 'penny' and 'penny' to 'coin'. The criterion proposed by Hirsch is then the following (p. 52): a sortal is merely a phase sortal, and does not name a real kind, if, when something ceases to fall under it, we continue to have an instance of another sortal to which the former sortal is subordinate. Say we have a red table, which we then paint green, making it the case that we no longer have an instance of the sortal 'red table'. We do, however, continue to have an instance of the sortal 'table', to which 'red table' is subordinate.

Therefore, we should not take 'red table' to name a kind, nor take the painting of a red table green as the ceasing to exist of an instance of this kind; instead 'red table' names a phase of things falling under the sortal 'table'. Clearly, by this criterion, 'penny-in-a-pocket' also comes out a phase sortal: taking such a penny out of its pocket will not obliterate an instance of the sortal 'penny', to which 'penny-in-a-pocket' is subordinate. Hirsch does not give his criterion a name of its own (instead referring to it as an addendum to another criterion he discusses); we shall refer to it here as the phase-substance criterion.

There is a difficulty with this criterion, however. It robs us of much more than just 'red table' and 'penny-in-a-pocket' as real kinds. An ontology in which everyday artefacts are accepted as real, as instances of real artefact kinds, must admit, in view of our own activity of making things out of other things and of our scientific knowledge of the composition of natural things, that new things can come into existence and that there are not just desks, bicycles and tigers but also lumps of matter, collections of cells, aggregates of molecules, and so forth. If we accept the existence of lumps of clay and also of statues that can be made out of such lumps as instances of real kinds (Gibbard 1975), then these two kinds of things show some independence and some dependence. A clay statue can come into existence while the career of the lump of clay has been underway for a considerable time, and can cease to exist, by being squashed, while the lump survives. Perhaps, *pace* Kripke, the statue does not even need this lump of clay to exist (as *this* statue) and could have been made out of another lump. However, a clay statue cannot exist without some lump of clay or other, whereas the lump can exist without it ever forming anything apart from a lump. Baker terms this relationship as one of *constitution* of one object by another: the lump constitutes the statue, but not the statue the lump.[2]

Say, then, that a particular lump of glass constitutes a glass vase. For the glass vase to go out of existence, it is not necessary for the lump of glass to be destroyed as a lump of glass; the vase could just be melted out of existence by giving the lump another shape. According to the above criterion, 'glass vase' cannot be an instance of a real kind; the only thing there, if any, is the lump of glass, and the vase is just a phase of it. In this way, the phase-substance criterion would rob of us of all things that are conceived as constituted, in Baker's sense, by underlying more 'primitive' or 'basic' things, as long as the constituted things are conceived in such a way that necessarily some restrictions on the material composition of the constituting thing hold. This is the case for 'glass vase', which can only be constituted by lumps of glass, and

[2]This usage, where one individual thing constitutes another, comes on top of the (more common) usage where a collection of cells or aggregate of molecules constitute a tiger at some particular moment (but at another moment the same tiger is constituted by another collection or aggregate) or the way an ensemble of components constitutes a bicycle (and at another time another ensemble may constitute the same bicycle). Evidently a broad theory of constitution like this has to contain, or be combined with, a theory of properties, because many of the properties that, say, a tiger has (like its weight or position) are not had by it independently of the collection of cells or aggregate of molecules by which it is constituted; it *inherits* these properties from its constituting object, as is the common expression.

also to 'human person', which must be constituted by human organisms. We cannot save glass vases by taking them to be something analogous to phases of the more encompassing kind 'vase', or human persons analogous to phases of the kind 'entity with mental states'. When a glass vase goes out of existence, necessarily a vase goes out of existence: the relation of 'glass vase' to 'vase' is as the relation of 'tiger' to 'animal', and if we accept tigers as forming a real kind next to animals, then why not glass vases next to vases? This can be blocked only if we reject both glass vases and vases as real kinds, which would mean that we end up with rejecting artefact kinds after all, and in the slipstream human persons as well. So it seems that an ontology that accepts glass vases as instances of real kinds constituted by lumps of glass, pennies as instances of real kinds constituted by flat pieces of metal and human persons as instances of real kinds constituted by human organisms must also accept red tables as instances of a real kind constituted by tables, pennies-in-a-pocket as instances of a real kind constituted by pennies, and so forth. And if, instead, red tables are conceived as phases of tables and pennies-in-a-pocket as phases of pennies, then so must pennies be conceived as phases of pieces of metal and persons as phases of organisms. That is, as long as no other criterion is available that allows us to distinguish between the former kinds of things and the latter.

The above arguments show that a seemingly innocuous criterion for setting restrictions on the reality of kinds in general can have a disturbing impact on what is accepted as ontologically respectful. We will again set this general criterion to work in presenting our view on how artefact kinds can be conceived of as real kinds. In the next section, we take the first step with a discussion of the second form of denial of the reality of artefact kinds mentioned at the beginning of Sect. 5.2, namely, a denial based on criteria that are related to the special character of artefacts.

5.4 Artefact Kinds and Natural Kinds

There are two major features of the way artefacts are generally conceived that are ontologically problematic. First, as already mentioned repeatedly, artefacts are seen as being dependent on human intentions in the sense that they come about through the actions of people, their designers and manufacturers, intended at producing things of precisely the constitution they eventually come to have. Second, and perhaps more controversial, even though they are the outcome of intentional efforts to make something like that, the precise material outcome is generally considered not to be what makes them the things they are. Artefact kinds are conceived of as things to which their precise material basis is immaterial, so to speak. It is necessary that they have an appropriate one, i.e. that they are constituted by some collection of matter that allows the performance of their function, but there is nothing what they are necessarily like. Artefact kinds are typically seen as being functionally defined: artefact kinds are equivalent to functional kinds. A clock is anything made by humans for indicating the time of the day, and a knife is anything made by humans to cut or to cut with. Neither kind fixes in any way what a clock or a knife should look like physically. Functional kinds are, as the saying goes, multiply realizable. Even if one considers it plausible that there are some constraints to the realizability of any particular functional kind, within these constraints the freedom for realizing the function is still large enough to bar any meaningful physical generalizations that all members of a functional kind answer to. As a result, artefacts belonging to one functional kind are widely divergent in every respect, except with respect to the function for which they were made. If an artefact's function is conceived as essentially intentional, then the two aspects are related, or even amount to the same thing.

To understand why these features must throw doubt on the ontological respectability of artefact kinds, let us compare artefacts to a contrast class of kinds of things that are unproblematically considered real by anyone who accepts a structured ontology, where we mean by this an ontology that posits various kinds of things that stand in mutual dependence relations to each other. The prime example of these unproblematically real kinds of things are natural kinds: kinds of animals, plants and minerals (by the latter we mean the broad category of 'geo-objects', subjects of the mineral kingdom, comprising gems, mountains, rivers, lakes, caves, and the like).[3]

To ask for an explanation of why animals and plants are unproblematically considered real would not be the right question to ask. Animals and plants are the reason why those philosophers who accept an ontologically structured universe do so in the first place. They are the paradigmatic examples of real kinds of things, which any adequate structured ontology should have to accommodate.

What unites the individual things into a natural kind is taken to be something intrinsic shared by all members. We believe that this is roughly correct, even though it may be very problematic to precisely characterize what is shared, and even though in biology further complications are introduced by the fact that individual organisms are causally united into species, which are subject to evolutionary change; see our Sect. 5.7 and the chapter by Reydon in this book for further thoughts on this. It is a consequence of their natural-kindhood that natural kinds are independent of human intentionality. The only dependence these natural kinds can be said to have is a dependence on the laws of nature. What criteria must whatever it is that unites different individual things into real, including natural kinds satisfy in order to be able to do this uniting? Wiggins (1980) expresses this by saying that the kind of thing something is, say an F, "determines (with or without the help of further empirical information about the class of Fs) (1) what can and cannot befall an x in the extension of F, and what changes x tolerates without there ceasing to exist such a thing as x, and (2) the relative importance or unimportance to the survival of x of

[3] Note that these natural kinds of things should be distinguished from the epistemological notion of natural kinds as used in the philosophy of science, basically as kinds allowing for inductive generalizations over their members. The latter concept is broader, including both stuff and kinds of things like animal and plant kinds; indeed the most often mentioned examples of natural kinds in the philosophy of science are stuff kinds like water and gold. For a discussion of the connection between the epistemological notion of a natural kind and the ontological notion of a natural kind of thing, see the contribution of Reydon to this volume.

various classes of changes befalling its compliants (e.g., how close they may bring x to actual extinction)", or alternatively, "determines either a principle of activity, a principle of functioning or a principle of operation for members of its extension" (pp. 68–70). Scientific research progressively reveals how this latter principle is realized, on the organic, cellular and molecular level. What it is to be an animal, or a tiger, is to be constituted in a way that determines how animals, or tigers, come into existence and in what various ways, under what various circumstances, they will continue to exist or can go out of existence.

In the case of artefacts, says Wiggins, their unity as a whole is not forced upon us through some internal principle of activity; instead, their unity is decided upon by us, depending on pragmatic considerations. Therefore, on Wiggins's view, with artefacts, we cannot be dealing with things that form a part of the 'furniture of the world'. A watch, for example, can be disassembled and then reassembled at any later moment and judged to remain in existence, or at least to be the same watch after reassembly as it was before disassembly, whatever befalls the separate components during the state of being disassembled. During reassembly some of its components could even be replaced by others and still we could judge that we are dealing with the career of one single persisting watch. But to what extent this is allowed is almost impossible to indicate in a general way: if during disassembly for cleaning just one component of the casing would be replaced bearing the initials of its first owner, we would (probably, depending on our relationship to this first owner) not consider it the same watch. Tracing the career of a single thing that is a watch can therefore not be done on the supposition that there is a thing that is a watch 'out there'; human considerations determine whether there is a watch there and which watch it is, and they do so ongoing, throughout the thing's career. This is exactly the sort of mind-dependence that instances of real kinds, including natural kinds, are not supposed to have.

On closer inspection, these arguments for fundamental differences in what unites instances of natural kinds and artefact kinds may not be so convincing. Making natural kinds the paradigm case of real kinds may blind us to a lot of indeterminacy concerning the intrinsic 'givenness' of many natural kinds. It may be obvious that animals come into existence, i.e. are born or hatch, at precise moments under precise circumstances; that they cease to exist, i.e. die, at precise moments; and that there are scores of circumstances that an animal will not survive. For plants, this is already less clear, not to speak of minerals. The well-known conundrum of the ship of Theseus, where the gradual replacement of the original boards by new ones, and the construction of a new ship out of the replaced boards, leads to the question which of the two ships is the same ship as the one that Theseus sailed, can now also be imagined for animals. In the near future we may be able to take out, over an extended period, an animal's organs one by one, replace them each time by other ones and use the original organs to create another animal that eventually consists entirely of the original animal's organs. Similar considerations apply to the distinction between substance and phase sortals, introduced in the previous section. Do rivers not change into valleys? It may be preposterous to claim that we will ever be able to change a tiger into a lion continuously, but it seems to us plausible that in the near future, with our increased knowledge of the biochemistry of DNA expression and cell metabolisms, we will be able to transform a virus or a bacterium of one kind into one of another kind such that always an organism is there. This would draw the carpet right underneath the real kindhood of viral kinds and bacterial species.[4]

A certain vagueness or indeterminacy, therefore, has to be accepted, even for the paradigmatic cases of natural kinds. Whether this compromises the very idea of characterizing real kinds, and thus the furniture of the world, in terms of natural kinds is an issue we leave open. Instead we look into the reasons for the indeterminateness of artefact kinds. What it means precisely to conceive of artefact kinds as functional kinds depends on the conception of function. Generally, two approaches are distinguished. One relates functions to an object's current properties, more in particular the physical capacities that enables the object to perform the function; a clock, for example, is then a physical thing that is used or can be used to tell the time, and a knife then a physical thing that is used or can be used for cutting with. The other relates functions to an object's history: a clock is anything made and designed to allow one to tell the time, and a knife anything designed and made to be used for cutting. In this latter conception of functional kinds, artefacts are explicitly mind-dependent, but not in an ongoing way. Neither approach, however, can serve to determine the identity of things sufficiently unambiguously, for the following reasons.

In either case, whether we take a function to be the purely historical property of being designed and made for a purpose or a material capacity allowing use for a purpose, the 'principle of unity or identity' we use has consequences that do not match our basic intuitions about artefact kinds. Suppose we witness the coming into existence of a tangible thing designed and made for the purpose of uncorking bottles. Then, as long as there remains a single tangible object there, any way it is modified or transformed, it remains the thing that it is: it remains this particular thing, i.e. tangible object, that came into existence in this particular way, through the intention to make something useful for a particular purpose. If it came into existence as a corkscrew, then as long as it remains a *thing* in the first place, it remains a corkscrew, even if it is run over by a train, hammered into a ball or transformed into a flute by relocation of its parts, to mention only a few possibilities, because its origin as a corkscrew is what identifies the thing.[5] The only additional criterion to restrict the thing's individuation as being an instance of this particular kind of artefact would be our decision to no longer treat it as that. But *that* would exactly introduce the sort of *ongoing* dependence on mental states that robs something of its

[4] See Soavi (2009a) for additional arguments in the same vein against a clear-cut distinction separating off natural kinds from their 'ontological surroundings'.

[5] Unless we see the running over, etcetera, as the intentional creation of a new object and the intentional annihilation of the old, functionally defined one, in the same way that our working at a reed stalk creates a flute *out of* a reed stalk. Note, however, that this requires the destruction of the original object as an extra criterion.

ontological respectfulness. The mere individuation through a generative event, therefore, is not sufficient for individuating a thing as an instance of a particular artefact kind.[6]

With regard to the alternative position of identifying a thing by its function, where a function is taken to be a capacity, the possibilities for modification without affecting the thing's identity are perhaps fewer but remain substantial. To mention only a very simple example, we could remove all the teeth of a fork but one, and the thing could still be used for spearing up food and so be a fork. A more serious difficulty for this position, however, is that it will not give us *artefact* kinds. Having the capacity for being used for some purpose does not discriminate between natural objects and artefactual objects. A glass splinter and a knife can both be used for cutting something. Many kinds, individuated in this way, will exclusively consist of natural objects: dogs for keeping flock of sheep together or for guarding property; cows, sheep, goats, horses, camels and llamas for being milked; and so forth.

To arrive at artefact kinds in this way, then, additional criteria are necessary, plausibly a reference to the object's origin, which would amount to a combination of the two approaches. Doing this neatly does not promise to be an easy task, however. A glass splinter, for example, is not a natural object but an artificial one, even though it was not made in order to be useful for cutting. Does or does it not belong to the *artefact* kind 'knife' or 'cutter'? But whichever way we do this, we end up with a counter-intuitive reduplication of kinds of things in this way, because introducing functional kinds will not rob us of natural kinds. As a consequence, we would have both a cow and a milk cow grazing in the meadow, both a dog and a sheep dog running around the flock and both a dog and a watch dog barking at the postman. Due to the well-known arguments grounding the notion of constitution for statues with respect to lumps of clay and for persons with respect to human organisms, these are separate, nonidentical things, because they differ in their modal properties: when a cow no longer gives milk, the milk cow goes out of existence but the cow remains, when a dog goes blind the sheep dog goes out of existence but the dog remains and when it loses its teeth the watch dog goes out of existence but the dog remains. This is not the ontology of common sense, whereas common sense was our starting point for accepting a rich, structured ontology including natural kinds.

In conclusion, as long as artefact kinds are conceived of as functional kinds, irrespective of whether the notion of function itself is conceived of in terms of human intentions or physical capacities, artefact kinds will suffer from an indeterminateness that throws their reality in doubt. In the next section, we will argue that a better route to the introduction of artefact kinds is, at least as a first step, by having their unity or identity determined through structural criteria, similar to natural kinds.

[6] This leaves out of consideration whose intentions would determine a thing's identity. In the literature about technical artefacts, some have questioned the default view that the prerogative for settling an artefact's identity rests with the designer and instead hold that 'social forces' may steer away an artefact's function from the one given to it by its designer. See, e.g., Preston (2006) and Scheele (2006). The latter view would still further complicate things; for one it reintroduces the artefact's ongoing ontological dependence on mental states.

5.5 Conceiving Artefact Kinds as Natural, i.e., Structural Kinds

Artefacts may be made to belong to functionally defined categories – cutting instrument and time-telling instrument – but each individual artefact is made by settling on a particular *design*, grounded in an *operational principle* to realize the intended function, and the design task may be considered complete, and the artefact kind 'individuated', when a blueprint giving all the relevant physical details is drawn up, and perhaps an indication of how the thing is to be manufactured (see Houkes and Vermaas's contribution to this volume). The blueprint determines a kind of thing, not an individual thing, and all things made on the basis of the same blueprint can be considered equivalent in all relevant aspects. Certainly all copies manufactured by the same procedure after the same blueprint can be expected to be structurally similar to a larger extent even than different tigers are similar to each other, not to speak of different mammals or different trees or different rivers. Functional kinds like 'cutter' (rather than 'knife', which already presupposes a particular operational principle') and 'clock' are, therefore, too broadly conceived to be examples of artefact kinds. Instead 'Zwilling J.A. Henckels Four Star 200 mm chef's knife' and 'Pasha Seatimer grand modèle automatique Cartier watch' should be considered as prime examples of artefact kinds. These are not functional kinds but kinds so defined that it is implied that their instances share a particular material, structure and configuration. As such, these instances satisfy law-like regularities of a strength equal to those satisfied by instances of natural kinds. Functional kinds arise by the grouping together of different artefact kinds.[7] Since in this grouping together the structural features of the artefact kinds play no role, or hardly plays a role, the broader functional kinds are indeed nominal kinds, unlike the biological case, where structural criteria do not just underlie (albeit in a causally complicated way) the identity of kinds at the level of species but also support (still in a causally complicated way) their being grouped into genera, families, orders, classes and phyla.

Wiggins seems to overlook the possibility of distinguishing between functional kinds and artefact kinds when he says that "the [functional] description [of an artefact] gives what it is usually impossible to specify in the other cases, an explicit nominal essence: whereas a finite and determinate set of marks suitable for definitional purposes is precisely not what the members of natural kinds endowed with a scientifically palpable real essence have in common." (1980, p. 87). A purely functional characterization gives us functional kinds like 'clock' and 'cutter', and these characterizations can be understood as "nominal essences" because they refer exclusively to ordering principles that exist in our minds, but exactly to that extent they lack any reference to structural properties and therefore to any *marks* suitable for definitional purposes. Contrary to what Wiggins suggests, such marks *are* in a sense available for natural kinds, in the form of specimens available not just for

[7] This view on the narrow character of artefact kinds has earlier been defended by Doepke (1987) and more recently Soavi (2009b), and by one of us in another context (Franssen 2013).

fixing the reference of our kind terms but also for specifying what sort of thing we are singling out.

Insofar as there are real things that artefact names refer to, these are to be conceived first of all as instances of structural kinds, on a par with natural kinds. This position is not so outrageous as it may at first instance seem. First of all, it could be objected that many things commonly seen as instances of artefact kinds simply cannot be structurally individuated. A simple example is a pair of earrings, which consists of detached components without a controlled causal link between the components. But we see no justification for the claim that it is the *pair* that is individuated as an artefact. Earrings are simply typically used in pairs, but their being what they are does not crucially depend on their existence in paired form. In many other cases, for example several distinct radio telescopes forming an array, there are controlled causal links, which definitely make the array into one thing.

Other prima facie objections against artefacts being instances of structural kinds can be likewise disposed of. With respect to kinds of matter, it is taken for granted that all samples of water are samples of one and the same matter kind 'water', in however way they came to be there, i.e. by having evaporated from a plant or by resulting from the intentional combustion of hydrogen with oxygen. Thanks to chemical and pharmaceutical industry, many kinds of matter exist only as the result of an intentional process of making them, but this does not make them ontologically special kinds of matter. Similarly, synthetic diamonds and rubies do not fall in another ontological category than natural diamonds and rubies, and if we create a sample of a synthetic mineral or element, like Americium, that does not occur in nature, it still falls into the class of minerals or elements.[8] The same would apply to engineered species, once we will be able to produce them.

If we do so conceive of artefacts, however, all basic characteristics of structural (natural) kinds should apply to them. One of the features for which this seems prima facie implausible is the fact that the meaning of natural-kind terms is extension-involving, a feature expressed in what is commonly called the Kripke-Putnam theory of reference. The meaning of the term 'water' is not a list of properties that water has, and that any sample that is a sample of water must therefore necessarily have, nor is the meaning of 'tiger' a list of properties that any tiger has, and therefore cannot lack without failing to be a tiger. Instead, 'water' and 'tiger' are defined as 'that kind of stuff' and 'that kind of thing', respectively, involving the pointing to a sample of water or a particular tiger. Now for artefacts, the situation is commonly thought to be entirely different. An artefact is considered to be defined by being something designed according to a plan, meant to be used in a particular way and supposed to match a precise description on the basis of which it can be established that it can be so used. More in particular, the suggestion is that an artefact kind is defined by the blueprint. Isn't the meaning of 'Pasha Seatimer grand modèle

[8] Silicon carbide (carborundum) and strontium titanate were long thought to be examples, but for both, naturally occurring minerals have been found, called moissanite and tausonite, respectively. Still, the naturally occurring samples were discovered long after the introduction of the synthetic form.

automatique Cartier watch' that it is the thing that has the material make-up indicated in the construction drawing of this watch type?

This, we think, would be mistaken. Instances of an artefact kind are typically *produced* starting from a blueprint, but that doesn't mean that an artefact kind should be *defined* as matching a blueprint, for the same reasons that we think this should not be done in the case of natural kinds. Suppose the design task for a particular new kind of artefact, say a biotransfaser, or more exactly a Smith & Watson multiswift type 2.0 biotransfaser, results in a blueprint, on the basis of which members of the kind are then produced. Suppose that once the artefact has been in use for some time, the first copies are sent in for maintenance or repair and that it is discovered that they do not match the blueprint exactly: either by accident or on purpose, modifications have been introduced that do not interfere negatively in an obvious way with what this specific biotransfaser is supposed to do. Should we then say that, contrary to what we all believed, there is no such thing as a Smith & Watson multiswift type 2.0 biotransfaser? No, instead, we will revise our ideas about Smith & Watson multiswift type 2.0 biotransfasers. They work (slightly) differently from the way we thought they worked. This exactly matches the way that our ideas about natural kinds are considered to be revisable. Phenomena like these in fact happen all the time in the design and manufacture phase of technology. Engineers discover properties of the artefacts they are busy designing just as much as these artefacts receive the properties that their designers have in mind for them. This is what goes on in testing and prototyping, and the reason why so much design is redesign. Therefore, for artefactual kinds of things, the process by which they come to be, including the blueprint supposed to describe them, *fixes the referent* for these kinds rather than *defines* them.

This argument is related to one that Putnam (1975) has developed: we could be wrong about everything we think is true for things categorized as artefacts, including that they are artefacts. This seems to be flatly contradicted by Thomasson (2007b), who claims that it is in the 'nature' of artefacts that we cannot be completely wrong about artefact kinds as we can be about natural kinds. This contradiction is only apparent, however. What Thomasson claims is that at least the designer of an artefact (kind) cannot be wrong about the artefact that has been designed and made so as to satisfy certain criteria, to put it very generally, because she knows she did so. But with respect to that artefact, everyone else can have it wrong, thinking, say, that that kind of thing grows on trees. And vice versa, if we are in fact not dealing with an artefact, and there is no designer/manufacturer, literally everyone can have it wrong and think that the things are artefacts, whereas they actually grow on trees (that this requires a conspiracy of some extent is irrelevant). So we can be wrong that things of a particular sort ('pencils' is Putnam's example) have the property of having been designed and manufactured to be used for some purpose, but if we are so wrong, we must also be wrong that these things are artefacts. Putnam's claim is a claim about the structural kind, and the historical properties of the members of this kind are among the properties about which we could be wrong; Thomasson's claim is a claim about a historical subclass of this structural kind, and insofar as these historical properties indeed single out a subclass we cannot be wrong about its members having these properties.

The conception of artefacts as instances of real, structural kinds may be a way of avoiding the indeterminateness problem, but it may come at the price of losing the distinction between artefact kinds and natural kinds. Let us see whether an appeal to the (intentional) history of things may solve this problem.

5.6 Artefact Kinds as Structural-Plus-Historical Kinds

Once we are able to arrive at sufficiently clearly individuated instances of artefact kinds on the basis of structural criteria, we can as a next step introduce the common-sense idea of artefact kinds through the inclusion of a historical criterion. The artefact kind 'Pasha Seatimer grand modèle automatique Cartier watch' then consists of those things that have all the structural characteristics of a Pasha Seatimer grand modèle automatique Cartier watch and that additionally have been designed and made to have this structure.[9] To such things, their history – their having been designed and made for some specific use – is by definition essential, as their redness is by definition essential to red tables if these are taken to form a real kind, and their being in a pocket is essential to pennies-in-a-pocket. Since the history of artefacts is partially articulated in terms of mental states – the aims, beliefs and decisions of designing engineers – this makes artefact kinds mind-dependent, but only in a historic or genetic sense, not in an ongoing or instantaneous sense, which is the sense usually meant when mind-dependence is stated to be straightforwardly at odds with real ontological status.

What is more, referring to our discussion of criteria for putting limits to what exists undertaken in Sects. 5.2 and 5.3, we may come to accept this historic form of mind-dependence as giving us ontologically real kinds. The phase-substance criterion introduced there has no quarrel with an ontological subdivision by historical properties. The relation of the kind Pasha Seatimer grand modèle automatique Cartier watch, all instances of which have the right intentional antecedents, to its parent kind made up by all objects having the structure of a Pasha Seatimer grand modèle automatique Cartier watch is equivalent to the relation that 'tiger' bears to 'animal', 'penny' bears to 'coin' and 'glass vase' bears to 'vase'. A Pasha Seatimer grand modèle automatique Cartier watch cannot go out of existence without a thing with the structure of a Pasha Seatimer grand modèle automatique Cartier watch going out of existence.

We also saw in Sect. 5.3, however, that the relevance of the criterion is highly questionable. It not only makes pennies-in-a-pocket mere phases of pennies, but also pennies themselves mere phases of pieces of metal. Judged by this criterion, any artefact that is distinguished only by its particular geometrical form from the underlying lump of matter, defined as a cohesive whole with a particular material composition, cannot be a thing of a real kind, but is just a phase of this underlying

[9] Note the similarity with Thomasson's characterization of artefacts quoted in Sect. 5.3.

lump. If we were to radically rearrange the components of a bicycle we create another phase of the underlying lump, no longer a bicycle but not ontologically different from, let alone inferior to, a bicycle. This can be blocked by denying to such lumps the status of individuals, for instance, by denying thinghood to things that can be split into smaller things of the same kind, but this would affect many natural kinds – rocks, amoebae, polyps and many plants – as well.[10] A further difficulty is that for many artefact kinds it is quite conceivable that a copy of one kind is transformed into a copy of another kind by a continuous process, i.e. such that a particular artefact of some broadly conceived kind is there all the time. By the criterion in question, we would then not be able to accept the narrow kinds as real kinds.

The phase-substance criterion was introduced in Sect. 5.3 for its ability to block a runaway proliferation of things and to secure a structured ontology containing only a limited number of unambiguously identified kinds. However, we fail to see how this can be achieved. We already discussed the difficulties surrounding red tables and pennies-in-a-pocket. Likewise, the notions of structurally defined natural kinds and artefacts as structurally plus historically defined kinds allow for a proliferation to which no easy limits can be set. As far as animals and plants are concerned, there is no reason why the species level should be the most basic one. We can introduce subkinds by further specifying kindhood by reference to, most plausibly, genetic details. Ultimately, in terms of the full DNA sequence of the ovum from which it grew, every individual animal in existence can be considered as being 'of its own kind', that is, a kind which happens to be instantiated by just one object, although it could still have indefinitely many instances, in line with the metaphysical necessity that kinds allow for multiple membership, as is shown, for animals, by the possibility of cloning (plants require less sophisticated methods). Similarly for historically defined artefact kinds, we can further subdivide such kinds, however finely detailed structurally, by continuing to specify details of a thing's design and production history. In this way we would also end up with any particular artefact being of its own kind. For either case, no criterion seems in sight to single out some specifications as ontologically respectable and others not. It seems, then, that there is not much to choose: to accept a structured ontology is inevitably to accept a maximally rich structured ontology.[11] Some aspects of this ontological structure are more salient than other aspects, figuring in our explanatory strategies and practical schemes, which indicates that ontology is not to be treated in isolation from epistemic and pragmatic concerns.

But even if with respect to the problem of proliferation, artefact kinds and natural kinds are in the same boat, artefact kinds may be considered ontologically suspect when they are defined in terms of historical, in addition to structural, criteria, and the more so if these criteria refer essentially to mental states. In the next section we argue that even in this respect the differences with natural kinds are not as substantial as they may seem.

---

[10] Cf. Wiggins remark on the difficulties presented by 'branching', to which he does not present a general answer (1980, p. 71).

[11] Sosa (1987, 1993) has argued, partly in a different context, for a similar conclusion.

5.7 Artefact Kinds, Biological Species and a Division of Explanatory Labour

It may be thought that artefact kinds are unique in being characterized by a combination of structural and historical features. However, we wish finally to emphasize that we are already acquainted with a similar situation with regard to natural kinds: it occurs also in biology. There we can also distinguish between tigers as forming a natural kind, structurally conceived, and tigers as forming a species, historically conceived. These two concepts are quite different. The natural kind 'tiger' is characterized structurally by the way tigers are made up of specific interconnected organs, which are made up of specific sorts of cells, and the metabolic processes in these cells and by the way they are governed by the DNA present in them and by the way parts of this DNA are expressed throughout the phases in the life of a tiger (to put it briefly). For discovering new facts about how a tiger works through observing a particular tiger, it is immaterial how that tiger came to be, whether we caught it in the wild or generated it in a laboratory, as long as it is structurally a tiger. Even a 'swamp tiger', a tiger that suddenly materialized in front of us by some process that we would be completely in the dark about, would do for this: everything we would discover about the behaviour of this tiger would be valid for all tigers. That is implied by the swamp tiger being a tiger.[12] In contrast, the species *Panthera tigris* is not a universal but an individual thing, a character in the history of life on Earth. This is now the common view of biologists (although of course, as any view in science, not uncontested; see (Reydon 2005) and the contribution of Reydon to this volume). A species is a thing consisting of all the individual species animals, causally connected by the generative links typical for the species. In mammalian species, for example, each animal is the causal outcome of a particular interaction between two species members, one male and one female. Actually, if we reserve the notion of 'membership' for the relation that an individual thing has to the kind it belongs to, we should, for reasons of conceptual clarity, choose a different notion for what relates an individual animal to its species. This relation is a form of parthood, or rather 'componenthood', similar to the way an animal's heart (if it has one) is a part, or rather component, of its body.

The relation between the natural kind 'tiger' and the species *Panthera tigris* is contingent. Not every member of the kind tiger belongs to *Panthera tigris*. Swamp tigers do not, for example, but neither does a tiger that has been engineered completely in the laboratory, including its DNA and the ovum from which it grew (assuming that that is the way it was generated), and that is kept apart from other tigers. Vice versa, it seems that every animal belonging to *Panthera tigris* is a member of the kind 'tiger', but this may not be so clear-cut. Species evolve continuously, and this creates difficulties for establishing what natural kind of thing a species is, similar to

[12] A caveat is in order here: the tiger's social behaviour with respect to other tigers and other animals could be deviant, since too little is fixed in this imaginary case about the tiger's learning and conditioning history. But this would not disqualify the animal as a tiger.

difficulties discussed in this chapter. We cannot go into these in more detail here, however. What is important to notice is that both concepts do explanatory work in biology, but different work. The structurally conceived kinds form the basis for explanations how organisms 'work'; how their behaviour, the properties they have as an individual organism, are realized through underlying causal mechanisms; and what the effects will be of interfering in specific ways with these mechanisms. The historically (i.e. in biology, evolutionary) conceived species forms the basis for explaining how the members of the kind came to have this structure and to contain these mechanisms. This explanation is available only insofar as the kind members are part of this historical individual; it does not help in explaining why a swamp tiger has the structure it has, or an engineered tiger. In the former case, we simply don't have a clue; in the latter case, it is the engineer's decision to make a tiger, and the decisions at every occasion where he or she could leave out or modify some aspect of tigerhood during the 'manufacture process' not to do so, and to continue the animal tiger-wise, that explains why the engineered tiger has the properties it has.

As the case of engineered tigers already indicates, we have a similar division of explanatory labour in technology.[13] The structurally conceived kinds form the basis for explanations how artefacts 'work'; how their behaviour, the properties they have as an artefact conceived of as a physical structure, are realized through underlying causal mechanisms; and what the effects will be of interfering in specific ways with these mechanisms. For this, the history of the particular artefact that represents the kind is irrelevant, as long as it is, on the basis of its structure, an instance of the kind. To explain how there came to be things with these structural properties, we need the historically conceived type of thing. This type will typically be thought of as a type characterized by a history that starts with a design task defined by a set of functional requirements. This, however, is a very impoverished view for explanatory purposes, since it leaves everything open about the 'dynamics' that link a resulting structural kind to a design effort defined in the intentional terms of functional requirements. For explanatory purposes, the idea of historically defined artefact types needs to be supplemented minimally by facts about engineering standards and methods, and how these operate and evolve. One way to do this, a social-scientific enterprise basically, could be to follow the model of biology and introduce something analogous to species, say 'cultures', being individuals that consist of humans and the tools they make and use. The ties that link them need not even be necessarily seen as intentional, that is, as explicitly referring to mental states; for explanatory purposes some reference to social *practices* of use and production may do just as well.[14] This cannot be elaborated here, however. What matters here is that much of what is considered unique to artefacts, both from an ontological and an epistemological perspective, can be pointed out to arise in a related form with respect to natural things.

---

[13] Note that the division between the two dominant conceptions of functions, one conceiving of functions in terms of (ahistorical) capacities and the other in terms of historically conditioned properties, corresponds with this division of explanatory labour.

[14] A position that has most recently been advocated by Preston (2012).

5.8 Conclusions and Discussion

Any defence of the reality of artefact kinds takes it for granted that the world is ontologically structured, containing various kinds of things connected by relations of mutual dependence, as we have, for example, for tigers and animals, or rubies and gems. We have seen that within such a structured ontology, arguments may be developed that the only real kinds are defined by their material structure, i.e. natural kinds, but these arguments are insufficient to brand artefact kinds as unreal. When specified in sufficient detail, artefacts may be taken to be instances of structural kinds and artefact kinds as structural-plus-historical kinds. Most natural kinds happen to come into existence by natural causes, but by our technological means we are now able to bring many of them into existence by artificial means. Likewise, most artefacts come into existence through pathways driven by human intentionality, but we could accept that they could come into existence in other ways as well. In neither case are we forced to judge this difference as ontologically relevant. If we choose to bring in the history of things to add ontological fine structure, again we have been unable to come up with arguments that can be used to brand the ensuing kinds as ontologically suspect, irrespective of whether a thing's history is specified in intentional terms. But it can be shown that any way in which some particular kinds of things, dear to common sense, are introduced into a structured ontology will open the way to many more things not so dear to common sense. Again, this seems not to depend on intentions playing a role in the ontological criterion; natural kinds allow for a potentially boundless fine structure just as much as kinds referring to historical or intentional aspects. Given the acceptance of an ontologically structured universe in the first place, it seems it must be accepted that this universe will be densely populated. If this is considered to be problematic, then perhaps it is the acceptance of an ontologically structured universe itself that should be reconsidered. However, it can be questioned whether in this respect anything is at stake at all. What is accounted for by things in a structured ontology is accounted for by properties in a flat one. Thomasson, who seems to us exceptional in openly favouring a maximally rich ontology, suggests that the assertions in the language belonging to her view, or at least the assertions that we care about, can be translated into 'equivalent' assertions in a language belonging to a sparser ontology (2007a, pp. 195–199).[15] If this is true, and we see no reason to expect that it is not, then it becomes difficult to see the point of any arguments pro the reality of this kind of thing and contra the reality of that kind of thing. The 'really' interesting questions left are conceptual and epistemological ones.

---

[15] See as well her (2008). Hirsch (1982, 2005) defends a similar, 'relativist' position vis-à-vis the correctness of any particular ontology.

References

Baker, L. R. (2007). *The metaphysics of everyday life*. Cambridge: Cambridge University Press.
Doepke, F. (1987). The structure of persons and artifacts. *Ratio, 29*, 36–52.
Franssen, M. (2013). The goodness and kindhood of artefacts. In M. J. de Vries, S. O. Hansson, & A. W. M. Meijers (Eds.), *Norms in technology* (pp. 155–169). Dordrecht: Springer.
Gibbard, A. (1975). Contingent identity. *Journal of Philosophical Logic, 4*, 187–221.
Hirsch, E. (1982). *The concept of identity*. Oxford/New York: Oxford University Press.
Hirsch, E. (2005). Physical-object ontology, verbal disputes, and common sense. *Philosophy and Phenomenological Research, 70*, 67–97.
Preston, B. (2006). Social context and artefact function. *Studies in History and Philosophy of Science, 37*, 37–41.
Preston, B. (2012). *A philosophy of material culture: Action, function, and mind*. New York/Milton Park: Routledge.
Putnam, H. (1975). The meaning of meaning. In K. Gunderson (Ed.), *Language, mind, and knowledge* (Minnesota Studies in the Philosophy of Science, Vol. 7, pp. 131–193). Minneapolis: University of Minnesota Press.
Reydon, T. A. C. (2005). On the nature of the species problem and the four meanings of species. *Studies in History and Philosophy of Biological and Biomedical Sciences, 36*, 135–158.
Scheele, M. (2006). Function and use of technical artefacts: Social conditions of function ascription. *Studies in History and Philosophy of Science, 37*, 23–36.
Soavi, M. (2009a). Antirealism and artefact kinds. *Technè, 13*, 98–107.
Soavi, M. (2009b). Realism and artefact kinds. In U. Krohs & P. Kroes (Eds.), *Functions in biological and artificial worlds: Comparative philosophical perspectives* (pp. 185–202). Cambridge, MA: MIT Press.
Sosa, E. (1987). Subjects among other things. *Philosophical Perspectives, 1*, 155–187.
Sosa, E. (1993). Putnam's pragmatic realism. *The Journal of Philosophy, 90*, 605–626.
Thomasson, A. L. (2003). Realism and human kinds. *Philosophy and Phenomenological Research, 67*, 580–609.
Thomasson, A. L. (2007a). *Ordinary objects*. Oxford/New York: Oxford University Press.
Thomasson, A. L. (2007b). Artifacts and human concepts. In E. Margolis & S. Laurence (Eds.), *Creations of the mind: Theories of artifacts and their representation* (pp. 52–73). Oxford/New York: Oxford University Press.
Thomasson, A. (2008). Existence questions. *Philosophical Studies, 141*, 63–78.
Wiggins, D. (1980). *Sameness and substance*. Oxford: Blackwell.

Chapter 6
Artifact Kinds, Identity Criteria, and Logical Adequacy

Massimiliano Carrara, Silvia Gaio, and Marzia Soavi

Abstract An important aspect of the alleged logical inadequacy of identity criteria for artifact kinds is the lack of transitivity. This formal problem can be eliminated through the use of a tool allowing for the gradual approximation of inadequate but normally adopted identity criteria. Although this does not solve the problem of so-called ontological respectability of artifacts, it shows that the mere lack of transitivity cannot be used to argue that artifacts are not ontologically respectable.

Keywords Antirealism on artifact kinds • Artifact kinds • Formal constraints on identity criteria • Identity criteria • Logical adequacy of identity criteria

6.1 Introduction

In the ontological debate on *realism*, it is possible to distinguish between *strong* and *weak realism* (for an overview on these distinctions, see Devitt (1991)). *Weak ontological realism* is characterized by the following thesis (considered the minimal antiskeptical thesis):

(WOR) There is a mind-independent world.

(WOR) can be combined with either skepticism or agnosticism about the possibility of knowing something concerning the nature of the mind-independent world, or, with the adoption of a positive epistemic stance, about our capacity for knowing the constitution and structure of the world. This second case leads to strong

M. Carrara (✉) • S. Gaio • M. Soavi
FISPPA Department, Section of Philosophy – University of Padua,
P.zza Capitaniato 3, 35139 Padova, Italy
e-mail: massimiliano.carrara@unipd.it; silvia.gaio@unipd.it; marzia.soavi@unipd.it

M. Franssen et al. (eds.), *Artefact Kinds: Ontology and the Human-Made World*, Synthese Library 365, DOI 10.1007/978-3-319-00801-1_6,

ontological realism (SOR). (SOR) can be further specified into two different forms by the following two theses:

(SOR1) The real world is an unstructured, indistinct blob of matter.
(SOR2) The world is structured: there are distinct objects, properties, etc.

In the first perspective (SOR1), there is no question as to which of the objects we individuate are real components of the world and which are mere projections of our thoughts: everything is a mere projection of our thoughts. On the other hand, if we adopt the second perspective (SOR2), we have the problem of selecting, from among the many entities such as objects, properties, events, facts, etc., the *real entities* – those really existing independently of our mental states. Hence, with respect to objects of a specific kind, one can be a realist, if one takes them to be real entities, or an antirealist, if one takes them to be mere projections of one's thoughts. That is, *ontological realism* and *antirealism* are stated with respect to the real existence of objects of a particular kind, not with respect to the real existence of the world.

In this chapter we confine our attention to a certain kind of object: artifacts. Adopting a different jargon, we can say that the problem for strong ontological realists of the second sort (SOR2) is selecting those objects that have *ontological respectability*. One standard (Quinian) solution in analytic philosophy is to argue that *identity criteria* are required for *ontological respectability*: entities are ontologically respectable, i.e., acceptable, if and only if they have clearly determined identity criteria. Think, for example, of the case of properties: following Quine, properties would not be ontologically acceptable because they do not have a suitable identity criterion. Applying the Quinian criterion to artifacts, it follows that artifacts are ontologically respectable if and only if they have clearly determined identity criteria. Antirealism with respect to objects belonging to artifact kinds has been defended by the argument that identity criteria for artifacts are *too weak to allow for their individuation*.

In general the following question with regard to identity criteria poses itself:

Question:
Are there general constraints to identity criteria for the individuation of real entities?

We distinguish between two kinds of constraints: *formal constraints* and *metaphysical constraints*.

Metaphysical constraints normally derive from the theses of the general framework adopted, for example, absolute identity versus relative identity or four-dimensionalism versus three-dimensionalism. Consider a philosopher who believes in relative identity. He thinks that identity is always relative to a general term and that the notion of absolute identity has to be abandoned and replaced by a multiplicity of relative identity relations: there is no absolute relation of identity between objects; no object is absolutely identical or distinct from another object, because there is no such relation as being just "the same" (see Geach 1962, p. 157). As a consequence, the absolute relation of identity needs to be replaced by a multitude of relative identity relations, each of them with a specific identity criterion. Again, consider the debate among four- and three-dimensionalists. The picture a three-dimensionalist has in mind is that a concrete continuant persists through time by *existing wholly and completely at each of several different times*. On the contrary, a four-dimensionalist argues against such a thesis. The two different ontological choices on concrete continuants give rise to different identity criteria, for example, for personal identity or for artifact identity.

Formal constraints on identity criteria are specified on the basis of the logical form of the identity criteria and some properties induced by it.

In the present work, we focus only on *formal constraints* or requirements on identity criteria; more specifically, we focus on a specific *formal constraint*: equivalence. Our aim is to argue that violation of this formal requirement is not sufficient to weaken the ontological respectability of entities. Because in most discussions on artifacts the failure of transitivity and, thus, of equivalence is assumed to be sufficient for antirealism, we consider our analysis particularly important for the case at issue.

6.2 Identity Criteria and Ontological Respectability

The introduction of the notion of *identity criteria* is usually attributed to Frege. According to Frege an identity criterion answers the following question:

Fregean question:
How can we know whether *a* is identical to *b*? (1884, §62)

In the literature, the Fregean question has been reformulated, relating identity criteria to kinds of objects (*K*) in at least three ways:

Epistemic question (EQ):
If *a* and *b* belong to *K*, how can we know that *a* is the same as *b*?

Ontological question (OQ):
If *a* and *b* belong to *K*, what is it for the object *a* to be identical to *b*?

Semantic question (SQ):
If *a* and *b* belong to *K*, when do *a* and *b* refer to the same object?

These three questions are related, respectively, to an epistemic, an ontological, and a semantic *function* of identity criteria.

To answer (EQ), we refer to conditions associated with a procedure for deciding identity questions concerning objects of some kind *K*. To answer (OQ), we refer to properties that objects of the same kind must share in order to be identical. In other words, identity criteria play a role in the identification process: to be able to identify objects (from both an ontological and an epistemic point of view), we need criteria that determine whether two items are the same object. Finally, an answer to (SQ) concerns sameness and difference of reference of simple or complex names. In our chapter we focus only on the *ontological* and *epistemic functions of identity criteria*.

6.3 Formal Constraints on Identity Criteria

Formally, an answer to (OQ) and (EQ) could be roughly represented by a binary predicate R, such that the resulting form of an identity criterion could be, for instance:

(IC) $\forall x \forall y \big((x, y \in K) \rightarrow (x = y \leftrightarrow R(x,y)) \big)$

or:

(IC′) $\forall x \forall y \big(x' = y' \leftrightarrow R(x,y) \big)$

where x' and y' are terms representing entities of the kind K suitably connected with x and y. Different ways of conceiving the form of an identity criterion are proposed in the literature[1]; here we consider just (IC). The informal reading of (IC) is the following: given two arbitrary objects, x and y belonging to a kind K, x is identical to y iff R holds between x and y.

Some supporters of (SOR2) (Lowe 1998; Wiggins 2001) argue that when we are able to answer (OQ) for K-objects by providing identity criteria that satisfy certain formal constraints, those objects are good candidates for being mind-independent objects or real constituents of the world. Two implicit assumptions in this realistic view are (1) that if a certain kind S is a real kind, then there are identity criteria for objects belonging to S and (2) that logical adequacy is a necessary constraint for identity criteria.

Those formal constraints concern the relation R in (IC), which is the identity condition of the criterion. In other words, given an identity statement $a = b$, R is a relation that holds between a and b. R is other than identity and analyzes what it is for the referents of a and b to be identical. In the literature some constraints on R have been just discussed (in Carrara and Giaretta 2001; Brand 1977; Lombard 1986; Wiggins 2001):

Non-vacuousness: the identity condition should not be vacuously satisfiable. Consider the following example (see Lombard 1986, pp. 32–33): let PO be the set of physical objects, S the set of relevant abstract objects, $R(x,y)$ the identity condition for PO, and $R'(x,y)$ the identity condition for S. Then let (IC) be defined as:

$$\forall x \forall y \Big(\big((x \in PO \vee x \in S) \wedge (y \in PO \vee y \in S) \big) \rightarrow \big(x = y \leftrightarrow (R(x,y) \vee R'(x,y)) \big) \Big).$$

The condition given above for the identity of x and y is not associated with a kind of entity in a metaphysically interesting sense, since the members of the alleged kind *physical objects or set* do not share any essential property. In a strong realist perspective, instead, the identity condition identifies as instances of the same kind those objects that share all the essential properties associated with that kind. Consider that, from such a perspective, the identity condition can be thought of as a

[1] For a survey, see Carrara and Giaretta (2004). See also Lowe (1989a, b, 1997) and Williamson (1991).

property of properties. Lombard calls such a property *determinable* since it determines a class of properties, or *determinates*, having that property. An example of a determinable is "being a spatiotemporal property," which can be considered a good candidate for an identity criterion for objects: if o and o' are physical objects, then they are identical iff they are alike with respect to all the properties that are spatiotemporal properties. A criterion of identity for K-objects, to be acceptable, must provide a determinable such that it makes non-vacuously sense to attribute determinates falling under the determinable to each K-object.

Informativeness: R should contribute to specifying the nature of the kind K of objects for which R acts as an identity condition. The identity condition does not completely characterize the nature of instances of K: to decide identity questions concerning items of K, we need the concept of K, which is not provided by the identity criteria. Nevertheless, an identity criterion should specify some nontrivial essential properties of objects of kind K. This means that the form of the relation cannot be tautological, for instance, of the form:

$$S(x, y) \vee \neg S(x, y), \text{where } S(x, y) \text{ is an arbitrary binary predicate.}$$

Partial Exclusivity: an identity condition for a kind K of objects cannot be so general that it can be applied to other kinds of objects. The example provided by Lombard is the following:

If x and y are both nonphysical objects, then x and y are identical iff they have the same individual essence.

Now, properties falling under the wide property "having an individual essence" do not apply only to nonphysical objects and can be part of the identity conditions for many kinds of objects. Living beings, for example, instantiate properties usually considered as individual essences, *being an individual x generated by gametes y and z*, for example, but they are not nonphysical objects.

Minimality: the identity condition for K-objects is required to specify the smallest number of determinables such that the determinates falling under them turn out to be necessary and sufficient to ensure identity between two objects of kind K. The determinables specified in the identity condition cannot be superfluous. Consider the following example (Lombard 1986, p. 38):

If x and y are both sets, then x and y are identical iff they have the same members and are liked by the same people.

The above criterion suggests that it is part of the very idea of sets that they are liked by people. But, clearly, it is not so. In order to rule out such cases, the formal requirement of *minimality* for identity criteria is introduced.

Noncircularity: the identity condition for K-objects cannot make use of the concept of identity itself; otherwise, it is circular. There has been a long debate about the circularity of (IC). Consider the criterion of identity for events proposed by Davidson (1969):

If x and y are events, then x is identical to y iff x and y have the same causes and effects.

Formally:

$$\forall x \forall y\big(x = y \leftrightarrow \forall z\big((z \text{ causes } x \leftrightarrow z \text{ causes } y) \wedge (x \text{ causes } z \leftrightarrow y \text{ causes } z)\big)\big).$$

One could argue that the above formulation is not formally circular, since the identity predicate does not occur in the right part of the biconditional. However, it has been observed that whether an event *e* has the same causes and the same effects as an event e' can depend on the solution to an identity question concerning entities of the same kind. On the right side of the biconditional, causes and effects are mentioned; since those are considered to be events, the identity criterion for events turns out to involve identity between events. In fact, to determine whether two events are the same, one is first required to determine the identity of the events taken as their causes or effects. One can thus conclude that identity is already presupposed.

This criticism corresponds to the denial that it is possible to give an explicative criterion of identity for objects of a certain kind, such as *events*. In fact, the formulation of such a criterion would involve a quantification over *all* objects for which the criterion is specified, and quantification presupposes the determinacy of the identity of the objects quantified (see on this Kripke 1978; Lowe 1989a, b; Quine 1985).

Non-totality: given at least two objects belonging to some kind *K*, *R* cannot be a property that every two *K*-objects share. Formally:

(C1) $R \subset K \times K.$

(C1) says that the relation *R* is a proper subset of the set $K \times K$: that is, there is some pair of *K*-objects such that the objects of the pair are not in the extension of *R*.

K-Maximality: *R* must be maximal with respect to *K*. In other words, *R* is required to be the widest dyadic property that makes an identity condition true.

A dyadic property *G* is wider than a property G' iff for any *x* and *y*, if $G'(x, y)$ then $G(x, y)$, but not vice versa.

In other words, the ordered pairs of G' are a subset of the set of ordered pairs of *G*. In such a way we always obtain a condition for an *ultimate kind* or *ultimate sortal* (*concept*) *K* (here, for the sake of simplicity, we use the term "kind" and "sortal" as synonymous. For a brief introduction to *sortals* in a realistic frame, see the next section).

The reason for introducing the formal constraint of *K-maximality* is this. Consider what Wiggins calls "a structure comprising only sortals" where sortals or kinds stand in relation to one another and have common members. Take two *sortals* C(1) and C(2). *Case 1*: Neither C(1) nor C(2) is a restriction of any other sortal and each is an *ultimate sortal*. If they have common members, then, "because they will cover identities relating to these common members [...] C(1) must be identical with C(2) or extensionally equivalent to it" (Wiggins 2001, p. 67). *Case 2*: Either C(1) or C(2) is an *ultimate sortal* and the other is not. In this case, if there are common members, the non-ultimate one gives a restriction to the other. *Case 3:* C(1) and C(2) have common members but no concept subsumes the other. In this case we have cross-classification, but some *ultimate sortal* will subsume both C(1) and C(2). "This picture of things," Wiggins argues, "founded in the nature of sortal[s] … and the absoluteness of identity, concedes everything that deserves to be conceded to the over-stringent demand that sortal[s] … should form a hierarchy" (Wiggins 2001, p. 67).

Uniqueness: R is unique with respect to K. This means that if there are relations $R_1, R_2, \ldots R_n$ such that (i) each R_i satisfies (IC) and (ii) each R_k is independent of each R_j (i.e., every R_k is neither narrower nor wider than each R_j), then at most one among the relations in $R_1, R_2, \ldots R_n$ provides a correct identity criterion for K-objects.

Equivalence: R must be an equivalence relation. On the left side of the biconditional in (IC), there is an identity relation that is an equivalence relation; consequently, the right side of the conditional is supposed to present an equivalence relation as well. R must then be reflexive, symmetric, and transitive.

Congruence: a is the same K as b iff the way in which a is K-related to b via R is sufficient for whatever is true of a to be true of b and for whatever is true of b to be true of a.[2]

6.4 Antirealism on Artifact Kinds

For the purposes of daily life, we are interested only in approximated procedures to individuate and identify objects. But from the perspective of a realist, most of these procedures do not correspond to identity criteria viable to be used in identifying objects that populate the outside world. Thus, most realists who assign to ontology the task of understanding the real structure and composition of the world turn out to defend antirealist theories with respect to some – even most – ordinary objects (for a general introduction to the topic, see Soavi (2009)). For example, Van Inwagen is an antirealist on artifacts but a realist for those organisms whose persistence is a *life* (1990, p. 145).[3] According to the *standard* ontological antirealist theory for artifacts, artifacts are objects associated with identity criteria that may be used perfectly for everyday practical purposes, but that do not have any serious ontological function.

In general, the antirealist ontological point of view on artifacts is based on Aristotle's idea that there is not a *real principle of unity* for them: in Aristotelian terms, artifacts do not have their own *nature* or *form*. Hence, they are not substances. The reason is that while changes in natural things happen by some specific principle of change and not only by chance, for artifacts we do not have such principles. Consider Aristotle's famous example of a bed: there are just principles of change governing the matter that composes it. But these are only incidentally principles of change for an artifact. Take the bed and suppose it is made of wood. Under certain

[2] We mention *congruence* in the list of constraints because, even if it is not a formal requirement on R like the others, it provides a further relevant criterion of selection among the possible R-relations that meet all the previous constraints.

[3] Another example of the same position is Wiggins (2001).

conditions it can burn, but such event depends on the material composing it not on the bed itself (on *artifacts* in Aristotle's metaphysics, see Katayama (1999)).

Following the Aristotelian point of view, antirealists with respect to artifacts argue that (A):

(P1) If a kind *S* is a real kind, then there are identity criteria for objects belonging to *S*.
(P2) It is not possible to have clear identity criteria for artifacts.

(C) Kinds of artifacts are not real kinds.

As an example of application of (A), consider David Wiggins's (2001, ch. 3) interesting view: he believes that – following (P1) – natural kinds are real kinds because they have identity criteria that isolate the essences of these kinds of objects via the determination of principles of activity for objects of natural kinds. In Wiggins's words, principles of activity are "law-like norms of starting to exist, existing, and ceasing to exist by reference to which questions of identity and persistence can be arbitrated" (2001, p. 83). Such principles describe regularities of the behavior of objects. The determination of natural kinds stands or falls with the existence of those lawlike principles. To better specify the nature of the principle of activity, Wiggins explicitly refers to some Aristotelian and Leibnizian passages:

> Things which exist by nature…such as animals and the organs of these or plants and the elementary stuff…have in them a principle of change or rest (in respect of place or growth and decline or alteration generally)…the nature of a thing being the source or cause of non-accidental change or rest…(Aristotle *Physics Book* II, Ch. I)

> [D]ivine law once established…has truly conferred upon [things] some created impression which endures within them, or… an internal law from which their actions and passions follow…if the law of God does in fact leave some vestige of him expressed in things…then it must be granted that there is a certain efficacy residing in things, a form of force such as we usually designate by the name of nature, from which the series of phenomena follows. (Leibniz 1969, p. 499 ff)

According to Wiggins we can think of an activity as a chain of internal and/or external causal interactions, a process determining the persistence of the object. So, for example, life or, more specifically, the metabolic processes of a human body are processes we can describe using lawlike sentences.

Wiggins' proposal on natural kinds resumes Putnam's characterization of natural kind terms: *x* is an *F* where "*F*" is a natural kind term if and only if, given good *exemplars* of the natural kind in question, the best theoretical description that emerged from inquiries into that kind would group *x* among the exemplars. Scientists or, more generally, experts make the characterizations of these good exemplars. They give us our best theoretical description. Ordinary speakers defer to experts the application of natural kind terms because experts know the criteria for the application of such terms (see Putnam 1975).

Wiggins' realism on natural kinds goes along with antirealism on artifact kinds (on Wiggins' conception of artifact kinds, see Soavi (2009) and Carrara and Vermaas (2009)). Following argument (A), Wiggins argues that artifactual kinds are not real kinds; the kinds to which artifacts belong are, for him, just conventional ones. Consider the following passage on artifacts:

> Artefacts identity does, however, present some difficult problems. **D**(iii) requires that the contribution that is made by an artefact-word such as 'clock' to the sense of '*x* coincides with *y* under the concept *clock*' should suffice to render this relation both an equivalence relation and a congruence relation. But how is this to be secured? And how are we to get for artefacts the effect that was got for natural things [...]? (Wiggins 2001, p. 91)

D(iii) is one of the requirements for the identity criteria associated with *sortals*, i.e., kinds associated with identity criteria viable for identifying objects that populate the outside world. Again:

> When the several points of discontinuous functioning, disassembly, and part replacement have been accommodated, the condition of coincidence and persistence for clocks that emerges is not merely weak. It is so undemanding that there seems to be nothing to prevent one clock [...] from clock-coinciding (*i.e.* coinciding in a manner supposedly sufficient for identity) with two distinct clocks. (Wiggins 2001, p. 92)

According to Wiggins, artifact kinds are functional kinds, which is to say that the objects falling under them are individuated on the basis of the functions they are supposed to perform. In other terms:

(FK) An object *o* belongs to a kind *S* iff *o* has the function *F*.

A pen is "any rigid ink-applying writing implement," a clock is "any time-keeping device," etc. It is part of our way of individuating artifacts to allow for a vast range of possibilities – parts replacement, dismantling, interruption of functioning, and other changes – to occur, leaving the identity of an artifact untouched. In order to fix a clock, for example, we may send it to a watchmaker, who may open it, replace its damaged parts, put the pieces back together, and let the very same clock start to function again: "[c]locks, for instance, may be made of a variety of different kinds of material and may function by radically different kinds of mechanisms" and "are collected up not by reference to a theoretically hypothesized inner constitution but under functional descriptions that have to be indifferent to specific constitution and particular mode of interaction with environment" (Wiggins 2001, p. 87).

As a result, the identity criteria that we associate with artifact kinds are not clearly determined and fail to meet the capital logical requirements of *equivalence* and *congruence*: either they are not transitive or they lead to contradictions. In the end, this is why objects belonging to artifact kinds are not ontologically respectable in our reading of Wiggins' perspective ((P2) of the argument (A)). Moreover, what is relevant for an antirealist position on artifacts is that all the available solutions for problems concerning identity of artifacts have a merely arbitrary or conventional *status*. In contrast, natural kinds are real kinds, in Wiggins' general realist frame, because they come with identity criteria isolating the essences of these kinds of objects.

In this chapter we consider a particular failure of identity criteria for artifacts with respect to the criteria we have mentioned, namely, the one that is most often discussed in the literature: non-transitivity. We wish to show that this failure alone is not sufficient to weaken the ontological respectability of artifacts, because it is possible to approximate non-transitive identity criteria in order to make them logically adequate.

6.5 Non-transitive Identity Criteria for Artifact Kinds

Let us consider some general identity criteria associated with all artifact kinds. First, we will consider a *material criterion*, then a *functional* one. For both cases we will demonstrate that the criterion is *formally inadequate*.

A *material criterion* for artifact kinds based on mereological essentialism – assuming the parts of a and b are arranged in the same way – may be the following:

(IC0) If a and b are K, then a is the same as b iff they have the same parts.

In (IC0) the identity condition for K-objects is given by the relation R_0 *having the same parts as*. It can be observed, though, that the *partial exclusivity* constraint is not met by R_0. However, for specific kinds, we can find a relation R that specifies R_0 and that meets the constraint. For example, we can say that if a and b are espresso coffee makers, then a is the same espresso coffee maker as b iff they have the same handle, the same knob, the same lid, the same pot, the same filter, the same funnel, the same rubber gasket, the same boiler, etc. For the sake of simplicity, though, we will stay at a general level when taking into account general relations like R_0, i.e., we do not specify a relation for each specific kind of artifacts like *handle*, *knob*, *lid*, *pot*, *filter*, *funnel*, *rubber gasket*, and *boiler*.

However, we do not usually use identity criteria like (IC0); according to our pre-theoretical intuitions, if we take a chair and replace only one part of it with a new part, we still have the *same* object.

A criterion closer to our intuition, then, could be the following:

(IC1) If a and b are K, then a is the same as b iff they differ at most in one part.

Nevertheless, even if we accept (IC1), some counterintuitive conclusions occur. If we accept that the replacement of one part of a chair does not change the identity of the chair itself, a *Sorites-like paradox* arises. Let $c_0, c_1, c_2, \ldots c_n$ be a series where c represents a chair and i the number of parts that have been replaced. c_0 is the chair at the initial stage and c_n the chair at the final stage, once all its parts have been replaced. According to our identity criterion, we state that c_0 is identical to c_1, c_1 is identical to c_2, and so on. Now, since identity is an equivalence relation, we obtain by transitivity that c_0 is identical to c_n – but we are not happy to accept that two objects with no parts in common are the same object.

A way to avoid this *Sorites-like paradox* is to restrict our identity criterion and say, for instance:

(IC2) If a and b are K, then a is the same as b iff they have at least 50 % + 1 parts in common.

Unfortunately, in this case too, the relation "have at least 50 % + 1 parts in common" fails to be transitive. Consider a chair c_0 and suppose that replacing a certain number of its parts with new ones yields a chair c_1 that has exactly 50 % + 1 parts in common with c_0. Now, replace some parts of c_1 with new ones and obtain c_2, which has exactly 50 % + 1 parts in common with c_1; it can happen that these parts are not the same parts that c_1 has in common with c_0. In general, then, we are not allowed to claim that c_0 is identical to c_2, because it is not true that c_0 has 50 % + 1 parts in common with c_2. *Transitivity* fails again.

Consider now an example of a *functional criterion* that has been proposed for artifacts.[4] Given an object a at time t and an object b at time t',

(IC3) If a and b are K, then a is the same as b iff there is a continuous material path between a and b and a and b are able to perform the *same function*.

The intuition behind (IC3) is that the identity of an artifact is strongly connected with its ability to perform a certain function. Of course, on the right side of (IC3), a criterion of material identity is also at work. Let us suppose, for the sake of simplicity, that this criterion poses no problems – the artifact is always of the same stuff – and just concentrate on the ability of the artifact to perform a certain function. Consider a case where an artifact gradually loses this ability.

Consider an object a that is able to perform some complex function F at time t_1 and gradually loses that ability after t_1, at the times t_2, t_3, … t_n. Let us suppose that, according to (IC3), $a@t_1=a@t_2$ and $a@t_2=a@t_3$ – where "$a@t_i$" stands for the object a at time t_i – and so on. Since the identity relation holding between each pair of elements in the succession is transitive, we are supposed to conclude that $a@t_1=a@t_n$. Nonetheless, because of the gradual loss of the ability to perform F, we are not inclined to accept that $a@t_1$ is able to perform the same function as $a@t_n$ and thus cannot conclude identity between $a@t_1$ and $a@t_n$.

For example, a laptop is an object that is able to perform a very complex function as the result of a combination of many subfunctions, such as those performed by the keyboard, monitor, DVD/CD reader and burner, battery, touchpad, etc. Let us suppose that the laptop at time t_1 is able to perform all those functions and that at time t_2 the *Alt* key no longer works due to minor damage. However, we still accept that $a@t_1$ and $a@t_2$ are able to perform the same overall function. Consider the possibility that a sequence of similar losses of functionality occurs until, at time t_n, the laptop is completely out of order. It is clear that $a@t_1$ is not able to perform the same overall function as $a@t_n$; thus, according to (IC3), we cannot say that $a@t_1$ is identical to $a@t_n$.

These examples of *material* and *functional criteria* show that the proposed identity criteria fail to be transitive. Despite their pre-theoretical plausibility, then, for formal reasons neither criterion is acceptable. Our proposal is to try to save these criteria by providing them with logical adequacy via a strategy of approximation.

6.6 Approximating Identity Criteria

In this section we consider two strategies of approximation for non-transitive identity criteria, one formulated by Williamson (1986, 1990) and the other by De Clercq and Horsten (2005). These proposals were not specifically developed for artifact

[4] Wiggins proposes (IC3) in (2001, pp. 72, 91).

identities (we have improved those approaches in Carrara and Gaio (2009a, b)); our aim is to adapt them to the case.

As we have seen in the examples above, it is common that the relations eligible to be identity conditions for artifacts are not transitive. However, it has been demonstrated that we can obtain equivalence relations that approximate those relations. Suppose one has a strong intuition about a certain relation R as a plausible identity condition for objects of some kind. Even if R turns out not to be transitive and therefore not a formally acceptable identity condition, it still seems to play a central role in the identifying process of K-objects. One might want to preserve one's intuition about the identity condition for K-objects as much as possible; one way to do so is to modify R in order to get another relation R', which is very much like R but transitive (see Williamson 1986, p. 381). Formally, R gives rise to a set of ordered pairs, not an equivalence class. One adds or removes pairs from the set given by R in order to obtain a unique equivalence class of ordered pairs. The set of ordered pairs modified in this way constitutes the equivalence approximation R'.

Consider one of the examples of inadequate identity criteria taken into account by Williamson (1986, p. 381) and De Clercq and Horsten (2005, pp. 371, 373), namely, an identity criterion for perceived colors. It seems plausible, when comparing the color of two or more different samples, to use an identity criterion like the following:

If a and b are perceived colors, then a is identical to b iff a and b are perceptually indistinguishable.

It is easy to verify that the relation "being perceptually indistinguishable from" is not transitive.[5] Imagine three monochromatic spots, a, b, and c: it might happen that one cannot distinguish the color of a from the color of b nor the color of b from the color of c, but that one can perceive a slight difference between the color of a and the color of c.

First, consider Williamson's proposal. Given a non-transitive R, let R_1, R_2, … R_n be equivalence relations that approximate R. Among them, we want to find the relation R_i that best approximates R. Williamson proposes to apply one of the following approaches:

Approach from above: Consider the smallest (unique) equivalence relation R^+ such that $R \subseteq R^+$.

Approach from below: Consider the largest (not unique) equivalence relation R^- such that $R^- \subseteq R$.

Let us clarify Williamson's suggestion by way of an example. Let D be a domain of objects:

$$D = \{a, b, c, d, e\}.$$

[5] On the topic of intransitivity of indistinguishability, see, for example, recent papers by Fults (2011), Raffman (2011), and Van Rooij (2011a, b).

Assume there is a candidate relation R, reflexive and symmetric, for the identity condition for the individuals of D, that means assume that:

1. For each element x of D, x is a relation R with x ($\forall x(Rxx)$).
2. If x is in relation R with y, then y is in relation R with x.

To simplify, when we define hereafter reflexive and symmetric relations on D, we will omit to write the pairs (a,a), (b,b), …. Moreover, if we write the pair (x,y), we omit to write the pair (y,x). Let R on D be the following:

$$R = \{(a,c),\ (a,d),\ (b,c),\ (b,d),\ (c,d),\ (d,e)\}.$$

R is not an equivalence relation because it fails to be transitive; for instance, R holds between a and d and between d and e, but not between a and e.

First, applying Williamson's approach from above, we obtain the smallest equivalence relation R^+ such that it is a superset of R:

$$R^+ = \{(a,b),\ (a,c),\ (a,d),\ (a,e),\ (b,c),\ (b,d),\ (b,e),\ (c,d),\ (c,e),\ (d,e)\}.$$

Applying the approach from below, we get a relation R^- that is not unique. For instance, one of the largest equivalence relations that is a subset of R is the following:

$$R^- = \{(b,c),\ (b,d),\ (c,d)\}.$$

We have obtained R^+ by adding four pairs to the initial relation R and R^- by removing three pairs from R.

De Clercq and Horsten define an approximation that modifies R as little as possible, via what they call the *overlapping approach*. An overlapping relation $R^\pm$ is neither a super- nor a sub-relation of R. Such a relation has the advantage of being closer to R than either R^+ or R^- is – that is, it modifies R less than either R^+ or R^- does. Applying the overlapping approach to Williamson's example, then, we obtain the following relation:

$$R^\pm = \{(a,b),\ (a,c),\ (a,d),\ (b,c),\ (b,d),\ (c,d)\}.$$

We obtain $R^\pm$ by adding one ordered pair to and removing another one from R. De Clercq and Horsten claim that $R^\pm$ is closer to R than either R^+ or R^- is. To understand why, consider how they determine which relation R_i among the approximations $R_1, R_2, \ldots R_n$ of a non-transitive relation R is the closest (or best) approximation with respect to R. First, they call *revision* any addition or removal of an ordered pair to or from R; second, they count the number of revisions made to get each approximation $R_1, R_2, \ldots R_n$ from R: such a number is called *degree of unfaithfulness*. They then state that a relation R_i is *the best approximation* with respect to R iff its *degree of unfaithfulness* is lower than that of all other approximations of R. With respect to the example above, we have seen that R^+ is obtained by adding four ordered pairs to

R, R^- by removing three ordered pairs, and $R^\pm$ by adding one ordered pair and removing another. Thus, the degree of unfaithfulness of R^+ is 4, the degree of unfaithfulness of R^- is 3, and the degree of unfaithfulness of $R^\pm$ is 2. The latter has the lowest degree of unfaithfulness; thus, $R^\pm$ is closer to R than are R^+ and R^-, because $R^\pm$ modifies R less than R^+ and R^- do.

According to De Clercq and Horsten, the fact that some identity criteria need "formal" improvement reflects the imperfection of commonsense concepts. Nevertheless, in spite of their imprecision, those concepts are useful for communication in everyday life. Imprecision in expressing identity criteria does not prevent us from applying the criteria themselves in ordinary circumstances. Even if identity criteria are not always fully adequate (for instance, if they fail to satisfy some logical requirements), they are sufficiently adequate for our usual, pragmatic purposes.

One could also ask what "best approximations" represent. Approximations are supposed to replace those non-transitive relations that are candidates for being identity conditions for some K-objects. It seems plausible to require identity conditions to be meaningful; nevertheless, the approximation process defined by Williamson and De Clercq and Horsten leads us to define relations that can be grasped only by enumerating the elements of their extensions and that are not intuitively captured by our concepts.

Moreover, we think that if an identity condition, containing a non-transitive relation R, is useful for communication in everyday life, its function is, at least, to give an answer to the epistemic question (EQ). Consider again, for example, the identity condition for perceived colors: as we have seen, we cannot precisely establish whether two items belonging to the kind "color" are identical. We rely on our perceptions which together with the assumption of transitivity lead to contradictions; thus, we cannot express an identity condition for perceived colors in a logically adequate way, nor can such an identity condition say exactly which colors are the same in reality. However, it is sufficient for our pragmatic purposes of comparing colors, and, so doing, it gives at least a partial answer to (EQ).

By contrast, an approximating relation is logically adequate, implying that it can determine whether or not two items are identical and say something about reality. Thus, it is plausible to think that an identity condition containing an approximating relation answers (OQ), even if such a relation pairs items in a way that does not correspond to our intuitions or that we cannot grasp with ordinary concepts. If so, an approximating relation cannot be useful for communication in everyday life and, thus, cannot have the function of answering (EQ).

6.7 Approximating Identity Criteria for Artifact Kinds

Let us adapt the sketched proposals to the case of artifacts. Consider, first, the material criterion of identity for artifact kinds:

(IC2) If a and b are K, then a is the same as b iff they have at least 50 % + 1 parts in common.

Let D be a domain of objects:

$$D = \{a, b, c, d, e\}.$$

Assume our R is "to have at least 50 % + 1 parts in common with." The relation is clearly reflexive and holds for all elements of D and is symmetric. As seen in Sect. 6.4, R is not an equivalence relation because it fails to be transitive.

Let R on D be the following:

$$R = \{(a,b), (b,c), (a,c), (c,d), (d,e)\}.$$

To clarify the example, assume that (IC2) applies to a specific kind of artifacts, say chairs, slightly modifying the example sketched in Sect. 6.5 above. The relation R chosen above, then, holds between chair a and chair b, b and c, a and c, c and d, and d and e, meaning that the elements of each pair have at least *50 % + 1 parts in common*. By contrast, R does not hold between chair a and chair d, a and e, b and d, b and e, c and e, as the elements of each of these pairs do not *have at least 50 % + 1 parts in common*.

First, applying Williamson's approach from above, we obtain the smallest equivalence relation R^+ such that it is a superset of R:

$$R^+ = \{(a,b), (a,c), (a,d), (a,e), (b,c), (b,d), (b,e), (c,d), (c,e), (d,e)\}.$$

In this way, all the chairs a to e come out identical.

Applying the approach from below, we get a new relation R^- that is not unique. For instance, one of the largest equivalence relations that is a subset of R is the following:

$$R^- = \{(a,b), (a,c), (b,c), (d,e)\}.$$

According to R^-, then, we are allowed only to say that chair a is identical to b and c, and chair d is identical to e.

We have obtained R^+ by adding five pairs to the initial relation R and R^- by removing one pair from R. R^- turns out to be the closest approximation of R, since its degree of unfaithfulness is 1. In this case, whatever approximation to R one takes, its degree of unfaithfulness cannot be lower. So, the overlapping approach provides a relation $R^\pm$ that coincides with R^-:

$$R^\pm = \{(a,b), (a,c), (b,c), (d,e)\}.$$

After all, in the example about chairs, it seems more acceptable to claim only that chair a is identical to b and c and that chair d is identical to e than to claim that all chairs are identical one to the other. For this reason, the approximation, granting only these identities, i.e., the one with the lowest degree of unfaithfulness, is *closer* to our intuitions about the relations among the elements of the domain.

The same train of thought can be applied in the case of the *functional criterion of identity* for artifact kinds.

6.8 Concluding Remarks

An important aspect of the supposed logical inadequacy of identity criteria for artifact kinds can be eliminated thanks to formal tools that allow for the gradual approximation to normally adopted identity criteria. This is not enough to solve the problem of the so-called ontological respectability of artifacts, but it shows that the mere lack of transitivity cannot be used to argue that artifacts are not ontologically respectable. We can solve the problem through the approximating tool illustrated in the chapter.

Even if approximation of the non-transitive identity criteria does not provide criteria for the ontological respectability of artifacts – logical adequacy is at most necessary but not sufficient for it – we can use it to show that arguments based on the supposed weakness of identity criteria for artifacts are not supported.

References

Brand, M. (1977). Identity conditions for events. *American Philosophical Quarterly, 14*, 329–337.

Carrara, M., & Gaio, S. (2009a). Approximating identity criteria. In V. A. Munz, K. Puhl, J. Wang (Eds.), *Language and world: Papers of the 32nd international Wittgenstein symposium*, Contributions of the Austrian Ludwig Wittgenstein Society, Vol. XVII, Austrian Ludwig Wittgenstein Society, Kirchberg am Wechsel (A), 2009: 127–133 (online on padua@research: http://paduaresearch.cab.unipd.it/2255)

Carrara, M., & Gaio, S. (2009b). On the logical adequacy of identity criteria. In J. Alcolea, et al. (Eds.), Actas del VI congreso de la sociedad de logica, metodologia y filosofia de la ciencia en espana, University of Valencia, Valencia 2009: 37–41 (online on padua@research: http://paduaresearch.cab.unipd.it/2267/)

Carrara, M., & Giaretta, P. (2001). Identity criteria and sortal concepts. In C. Welty & B. Smith (Eds.), *Formal ontology in information systems* (pp. 234–243). New York: ACM Press.

Carrara, M., & Giaretta, P. (2004). The many facets of identity criteria. *Dialectica, 58*, 221–232.

Carrara, M., & Vermaas, P. E. (2009). The fine-grained metaphysics of artifactual and biological functional kinds. *Synthese, 169*, 125–143.

Davidson, D. (1969). The individuation of events. In N. Rescher (Ed.), *Essays in honour of Carl G. Hempel* (pp. 216–234). Dordrecht: Reidel.

De Clercq, R., & Horsten, L. (2005). Closer. *Synthese, 146*(3), 371–393.

Devitt, M. (1991). *Realism and truth*. Oxford: Basil Blackwell.

Frege, G. (1884). *Grundlagen der Arithmetik. Eine Logisch- Mathematische Untersuchung über den Begriff der Zahl*. Breslau: Köbner.

Fults, S. (2011). Vagueness and scales. In P. Egré & N. Klinedinst (Eds.), *Vagueness and language use* (pp. 25–50). New York: Palgrave Macmillan.

Geach, P. T. (1962). *Reference and generality. An examination of some medieval and modern theories*. Ithaca: Cornell University Press.

Katayama, E. G. (1999). *Aristotle on artifacts: A metaphysical puzzle*. New York: State University of New York Press.

Kripke, S. (1978). *Time and identity*, manuscript.
Leibniz, W. (1969). On nature itself, or on the inherent force and actions of created things. In Id. *Philosophical papers and letters* (L. E. Loemker, Trans.). Dordrecht: Reidel.
Lombard, L. (1986). *Events: A metaphysical study*. London: Routledge & Kegan Paul Books.
Lowe, E. J. (1989a). Impredicative identity criteria and Davidson's criterion of event identity. *Analysis, 49*, 178–181.
Lowe, E. J. (1989b). What is a criterion of identity? *The Philosophical Quarterly, 39*, 1–21.
Lowe, E. J. (1997). Objects and criteria of identity. In B. Hale & C. Wright (Eds.), *A companion to the philosophy of language* (pp. 613–633). Oxford: Blackwell.
Lowe, E. J. (1998). *The possibility of metaphysics. Substance, identity, and time*. Oxford: Oxford University Press.
Putnam, H. (1975). The meaning of 'Meaning'. In Id. *Mind, language and reality: Philosophical papers* (pp. 215–271). Cambridge: Cambridge University Press.
Quine, W. V. O. (1985). Events and reification. In E. Lepore & B. McLaughlin (Eds.), *Actions and events: Perspectives on the philosophy of Donald Davidson* (pp. 162–171). Oxford: Blackwell.
Raffman, D. (2011). Vagueness and observationality. In G. Ronzitti (Ed.), *Vagueness. A guide* (pp. 107–122). Berlin: Springer.
Soavi, M. (2009). *Realism, antirealism and artefact kinds*. Padova: Cleup.
Van Inwagen, P. (1990). *Material beings*. Ithaca: Cornell University Press.
Van Rooij, R. (2011a). Vagueness and linguistics. In G. Ronzitti (Ed.), *Vagueness. A guide* (pp. 123–170). Berlin: Springer.
Van Rooij, R. (2011b). Implicit versus explicit comparatives. In P. Egré & N. Klinedinst (Eds.), *Vagueness and language use* (pp. 51–72). New York: Palgrave Macmillan.
Wiggins, D. (2001). *Sameness and substance renewed*. Cambridge: Cambridge University Press.
Williamson, T. (1986). Criteria of identity and the axiom of choice. *The Journal of Philosophy, 83*(7), 380–394.
Williamson, T. (1990). *Identity and discrimination*. Cambridge, MA: Blackwell.
Williamson, T. (1991). Fregean directions. *Analysis, 51*, 194–195.

Part II
Artefact Kinds and New Perspectives

Chapter 7
Creating Artifactual Kinds

Jesús Vega-Encabo and Diego Lawler

Abstract The aim of this chapter is to assess how two widespread types of theory on the nature of artifact kinds (i.e., functional and intentional theories) address the *creation requirement*, which demands an account of the appearance of genuinely new artifacts resulting from intentional creative processes. It attempts to show that both types of position do not satisfy this requirement. Functional theories that refer to a causal reproductive history cannot account for the nature of newly created artifact kinds, because artifacts belonging to these kinds do not have ancestors. Intentional theories that make the emergence of a new artifact kind dependent on the possession of a new concept of that artifact kind face a dilemma: either they have to excessively weaken the conditions for possessing a concept of an artifact kind or they need to concede that the constitution of the newly created kind cannot depend completely on such a concept. We conclude the chapter by arguing that if any account which treats artifacts as products of intentional creations cannot be separated from adopting a stance on what artifacts really are, then there are four aspects that must be taken into account for satisfactorily dealing with the ontology of artifacts.

Keywords Artifact concepts • Artifact normativity • Artifactual novelty • Creation requirement • Functional account • Historical-intentional account

---

J. Vega-Encabo (✉)
Facultad de Filosofía y Letras, Universidad Autónoma de Madrid,
Crta. Colmenar, km. 15, 4, Madrid 28028, Spain
e-mail: jesus.vega@uam.es

D. Lawler
Sociedad Argentina de Análisis Filosófico, Calle Bulnes 642,
1176SS – Ciudad Autónoma de Buenos Aires, Argentina
e-mail: diego_lawler@yahoo.com

M. Franssen et al. (eds.), *Artefact Kinds: Ontology and the Human-Made World*,
Synthese Library 365, DOI 10.1007/978-3-319-00801-1_7,

7.1 Introduction

Among those aspects that characterize our world of cultural objects and technical artifacts, one fact stands out: the sustained and continuous appearance of *new* kinds of artifacts. Of course, novelty can be said or elicited in multiple ways. In most cases, new artifacts bear strong resemblances to existing ones, being variations of previous artifacts, modified in their material or functional structure to adapt to new artifactual niches. But often artifacts are *genuinely new*, namely, original productions showing a very weak family resemblance with the population of existing artifacts or even lacking precedents of any kind in that population. In these cases, the creation of a novel artifact involves the appearance of a new kind of artifact that may present a new solution to an old problem or offer a new solution to a new problem. In many other cases, new artifacts are only *relatively new*, i.e., they are artifacts belonging to existing kinds but modified to varying degrees (modified in their material or functional structure, redesigned, adapted to new artifactual niches, etc.). These artifacts show strong family resemblances to certain artifacts which already exist, and they often represent alternatives to them.

Our aim in this chapter is to assess how certain theories about artifact kinds, which impose requirements that are specific to artifacts, address the *creation requirement*. This requirement demands an account of the appearance of genuinely new artifacts (and their corresponding kinds) resulting from intentional creative processes. We will show that two widespread types of theories on the nature of artifact kinds and artifacts, *functional* and *intentional* theories, do not appropriately satisfy this requirement. Broadly speaking, a functional theory is a theory that gives an account of the nature of artifact kinds and artifacts in terms of what they are supposed to do, i.e., in terms of the functions they perform. An intentional theory is one that reflects on the nature of artifact kinds and artifacts in terms of the intentions and concepts of their makers.[1] A critical examination of both theories will confront us with the following dilemma: (1) if the theorist adopts a functional theory and chooses to account for the ontology of artifacts in terms of proper functions and, in turn, to account for those functions by referring to a causal reproductive history, her explanation will exclude an account of the nature of newly created artifacts; (2) however, if she adopts an intentional theory and makes intentional dependence into a constitutive aspect of being this or that artifact, she must argue that the

[1] Although both theories can be combined in a single and complex one, we will treat them separately. Our purpose is not to delve into the philosophical advantages of having a unified account of the nature of artifacts; our interest lies rather on diagnosing how the intuitions that give rise to these two different theories face the challenge of meeting the creation requirement. In this sense, both theories play the role, so to speak, of ideal types in approaching questions regarding the nature of artifacts. As the vocabulary we are employing suggests, theories addressed in this chapter do attempt to account for the nature and identity of individual artifacts in terms of the primary kinds they belong to. A metaphysical explanation of artifact kinds has implications for delineating identity conditions of individual artifacts. And vice versa, a theory that specifies identity conditions for artifacts helps to see some significant points in developing an ontology of artifact kinds. Anyway, these issues are not directly addressed in this chapter.

emergence of a *new* artifact depends on the possession of a *new* concept of that artifact, a concept that guides the production of that sort of object. If this is so, then the intentional theorist faces a further dilemma: either she has to excessively weaken the conditions for possessing a concept of an artifact or she needs to concede that the constitution of the newly created kind *cannot* depend completely and effectively on the concept, because, especially in cases of creation, the creator cannot possess the concept that will guide the production of the new artifact.

In what follows, we will not be concerned with the philosophical roots of these dilemmas or delve into the reasons that make this problem persistent. We have a much more modest goal: to render the dilemma fully visible and argue for its philosophical plausibility. To this end, the chapter is divided into three main sections. The first section introduces what we call "the creation requirement" and argues that this requirement should be considered a *factum* that any philosophical theory on the nature of artifacts must be able to meet. The second section presents the first horn of the dilemma, the problems that functional theories as we conceive of them face when addressing the creation requirement. The third section discusses how historical-intentional accounts try to deal with this problem. It will become clear how this approach reaches an intractable dilemma that generates a kind of conceptual paralysis when facing the challenge of the *creation requirement.*

7.2 The Creation Requirement

The artifactual world is impressive in its novelty. A neophilic compulsion seems to characterize our contemporary culture,[2] in comparison to ancient times where artifactual lineages were more stable and display less variability. With varying frequency, *any* technical culture witnesses the emergence of new artifacts. But what makes the appearance of *new* artifacts possible? Humans organize matter in varied ways and under many different forms. Some ways of organizing matter are unprecedented and, under certain conditions, give rise to new objects. We can outline some criteria for grouping new objects under more or less defined categories or kinds. These criteria are based usually on functional principles; they answer a "what-for question" about this new object. For example, we identify an object as a corkscrew by specifying what the object is for, that is, for removing corks. We can then establish: "this new configuration of matter is a corkscrew." If there are other previous instances of this sort of object, then the novelty of this new arrangement of matter rests on the fact that it is a modified example of an already existing kind.

[2] Under the heading of "contemporary culture," many different cultures are included. Contemporary technological society is dominated by this neophilic compulsion. However, there are social groups whose particular cultures seem to be resistant to this neophilic compulsion, for example, the Mennonite culture. In this more traditional culture, artifactual lineages are more stable; the variability is less intense; nevertheless, they confront, though not so often, the same situation of new created artifacts.

This variation could concern various aspects, ranging from its materials to its principles of operation.

Nonetheless, an innovation could also be a radical novelty, in which case we are in the presence of an artifact that cannot be understood as a modification of an existing kind and has no relationship to the current set of artifacts.[3] We have identified a *factum* that requires explanation: *new* artifacts repeatedly appear in our world. A complete and plausible philosophical theory of artifacts and artifact kinds cannot ignore this fact of human culture. Taking this as a background condition, it is easy to see how it imposes a certain explanatory demand on theories concerning the ontological condition of new artifacts. A philosophical theory about the nature and identity conditions of artifacts and artifact kinds must apply to artifacts that cannot be identified as belonging to an existing kind because their very appearance marks the beginning of a new artifactual kind. Or to put it differently, any theory of this type must be able to account for the identity of genuinely novel artifacts. We call this the *novelty condition*.

This condition leads us to a second explanatory demand: we must account for the fact that a good portion of these genuinely new artifacts are the result of *our creation*. It is not enough to explain the appearance of novelties in a space of artifactual objects with their own identity conditions. It is also necessary to identify conditions that make the genuine creation of artifacts possible under constraints of intentional dependence. Therefore, the fact that they are *created* objects imposes an added requirement to the novelty condition, a requirement that underlines the strong reliance of genuinely new artifacts on human intentional activity. The creation of artifacts and artifact kinds is an achievement of intentional agents.[4] Both explanatory demands can be formulated in terms of the following creation requirement:

Creation Requirement (CR): Any theory about the nature of artifacts must offer an explanation of the ontological nature of genuinely new artifacts that are the result of our creation.

We offer a strong reading of this requirement because we are interested in the appearance of objects that embody a set of material and functional properties that

---

[3] A radical novelty is not an absolute novelty. Our judgment about what constitutes a radical novelty always presupposes a background of previous artifacts against which that radical novelty emerges. For a discussion of the incoherence of the notion of absolute novelty, see Briskman (1980, p. 95).

[4] Briskman (1980) discusses the conditions that a novel product (scientific or artistic) must satisfy to be a new creation. Although it is Briskman's aim to provide an answer to the question how creativity is possible, these conditions could be applied to characterize novel artifacts that are the result of human creative activity. Roughly speaking, we would have the following formulation: a novel artifact is a creative one if (a) it is valuable novelty (i.e., it solves a problem), (b) it conflicts with the background available solutions (tradition), and (c) it meets existing standards of acceptability (Briskman 1980, p. 97). This formulation can be challenged by noting that a genuine new artifact subverts condition (c); such an artifact engenders in a way new normative criteria. Condition (c) seems to be a conservative one, and it can be an obstacle to fully understand the appearance of genuine new artifacts. However, it is neither our aim to provide a theory of creativity nor to discuss Briskman's conditions.

cannot be identified in reference to lineages of existing artifacts (i.e., genuinely new artifacts).[5] *CR* reflects the fact that there are artifacts that inaugurate a new lineage, in the sense that no artifact of this kind existed before.[6] There were times at which corkscrews did not yet exist, and there are times at which corkscrews do exist. Does the emergence of a genuine novelty involve, ipso facto, the emergence of a new artifact kind? The appearance of the Newcomen steam engine is radically new. The introduction of a new operational principle (the use of steam in a cylinder and atmospheric pressure after bringing about a vacuum) expresses a genuine novelty. Not even the fact that the use of steam was a previously known principle lessens its novelty, since it is the final disposition of the machine that is original. We can all think of paradigmatic examples of innovations, outcomes of intentional creation that lack recognized lineages, such as the light bulb, the electric motor, or the microchip. Following this line of reasoning, it is certainly plausible to think that some sorts of technical development can only be described in terms of the emergence of a new kind of artifact.

So *CR* picks up the following fact: some genuinely new artifacts are the result of our creation. This fact has metaphysical relevance because the emergence of a genuine new artifact also means the introduction of a new kind.[7] Since newly created artifacts cannot be placed within actual lineages or understood by reference to these lineages, it seems impossible to determine their identity and persistence conditions in these terms. That is, we cannot establish their ontological nature by reference to the identity conditions of other artifacts. Newly created artifacts establish identity conditions for objects that are *added* to our ontology. Any adequate metaphysical theory about the ontological nature of artifacts must be able to account for the identity conditions of genuinely novel artifacts – which, by being genuinely novel, do not belong to any existing artifact kind.[8] *CR* claims that these newly created artifacts

[5] A weak version of the novelty condition suggests that we should understand the appearance of a new artifact that shows new properties in terms of a set of objects of the same lineage. These new properties may include new functional realizations, a different material structure, or a combination of both.

[6] We take *CR* as a requirement to be imposed on theories of artifact kinds that is specific to artifacts. Our reading of *CR* claims that any candidate theory on the nature of artifact kinds and individual artifacts that causes trouble for understanding novel artifacts, or that precludes the possibility of novel artifacts, or that makes them mysterious is flawed.

[7] Baker (2007) has defended a form of constitutivism that accounts for the coming into existence of both new artifacts and new primary kinds. She suggests that we can "understand ontological novelty as the evolution or introduction at some time or other of objects of new primary kinds – e.g., the first organisms or Galileo's first telescope" (p. 234), and she adds, "[a] new primary kind (natural or artifactual) is a genuine novelty whose evolution or introduction makes the world ontologically richer" (p. 234). Baker's view is articulated as a way of dealing with any kind of eliminative attempt in the field of everyday objects. Our discussion of the coming into existence of a new artifact assumes that the eliminative strategy is not a viable option on this point.

[8] It is obvious that novelty is a matter of degree and that probably our judgments of novelty are influenced by many different social and cultural factors that compose a background of recognition and a ground for adapting our taxonomies and practices of identifying kinds. Not all differences with previously existing artifacts constitute a genuine novelty in the sense that is meant in our argument, that is, as involving the need to identify the artifact under new conditions of appearance, persistence, and extinction. We are not committed here to the idea that these phenomena need to

need to be covered by our best ontological theory of artifacts and that this theory should be equally applied to standard artifacts (i.e., artifacts that do not represent any novelty) as well as to genuinely new artifacts that result from our creation. Therefore, any metaphysical account of artifacts must be able to explicate not only the creation of genuinely novel artifacts but also the ontology of existing artifact kinds and particularly in both cases their ontology with respect to how they come into being. In this sense, *CR* is an unavoidable challenge for any metaphysical theory about the nature of artifacts.[9]

7.3 New Artifacts and Functions

A new artifact must be characterized according to the identity conditions that are proper to it qua such an artifact and which constitute the kind to which it belongs. What relationship must exist between different individuals so they can be consistently grouped under one kind? Two responses are predominant in the literature on artifacts: you can identify the kind either by generalizing the traits of an individual with a certain function to other individuals with the same function or through the generalization of these traits to objects created with the same *intention* (Grandy 2007, p. 29). Functional theories assert that sameness of function is the relevant criterion for kind membership. Artifactual kinds are primarily functional kinds, and, thus, the creation of a new artifact entails the creation of a new functional kind. An adequate description of an artifactual object would consist in the identification of a proper functional kind to which it belongs. In this view, then, each new artifact is *new* in virtue of its (proper) function.

In this section, we will review how functional theories confront the challenge of *CR*. Firstly, we consider etiological theories; secondly, we will discuss Preston's pluralistic theory; and finally, we will outline how to meet *CR* if we assume that some version of intentional etiological theories is acceptable.

be understood in such strongly ontological terms; our intuition, simply put, claims that any metaphysics of artifacts and artifactual kinds has to take into account the fact that each newly created one is an element of the artifactual reality. Therefore, as part of its reflection on the nature of existing kinds, a metaphysical theory has to deal with newly created artifacts.

[9] In standard versions of theories about the nature of artifacts, providing an account of the nature of an artifact involves referring this artifact to a determinate kind. Among other things, this accounts for the fact that the identity and persistence conditions of artifacts are given by their belonging to an existing kind. We should make explicit our view on the *reality* of artifactual kinds: for us, this use of the terminology of "kinds" only reflects how different instances of a same kind can inductively provide information about other members of the same kind (i.e., the epistemological sense). We will not make any explicit commitment to the existence of *real* kinds of artifacts in our argument. This terminology is very slippery when applied to cultural and artifactual products of human making. How we usually categorize these objects is governed by criteria that are often in open conflict. It could be that there are only individual artifacts and our groupings in kinds are the result of conventional practices. Still, there are artifacts that are genuinely *new* and that is the fact that we do not want the reader to lose sight of – and this is sufficient for our argument to run.

The proper functions of an artifact (what the artifact does) are those functions that have been fixed through a history of selection that in a certain way fits the biological model of selection upon success. Etiological theories of function ascription fix proper functions by appealing to the causal history that led to the appearance of an artifact with the structure it effectively has. According to Vermaas and Houkes (2003), these theories can be classified along two axes: on the one hand, according to whether the etiology of the item is reproductive or not and, on the other hand, according to whether the etiology is intentionalist or not. Prima facie, reproductive etiological theories of an intentional character seem more appropriate for artifacts, as they are able to accommodate intentional factors linked to the existence and persistence of artifacts. In principle, it would be implausible to admit a theory of artifact functions in which the causal history that fixes the particular function did not involve – at least partially – intentions. Furthermore, efforts to apply a reproductive model based on biological organisms in the field of artifacts have always encountered insurmountable difficulties, especially when trying to explain what is involved in the processes of copying (Vermaas and Houkes 2003; Soavi 2009).

Etiological theories meet further obstacles when trying to account for the novelty condition. Reproductive etiological conceptions seek an explanation of the emergence of new artifacts by appealing to the notion of proper function. In general, there is artifactual novelty when an artifact presents a new proper function. The key problem here is to determine the relevant causal history of new artifacts that have no reproductive ancestors. According to this model, novelties can be expressed as variations that are often introduced by the intentional and creative intervention of the designers, and afterwards, various selection pressures come to determine the success or failure of the new trait or artifact. This selection process is what determines the proper function and identity of the artifact.

However, there is a central tension at the heart of this view. On the one hand, the theory claims that novelties should be understood as variations resulting from intentional and creative intervention. On the other hand, it argues that the proper function is what gives identity and persistence to the artifact. But, if novelty emerges as the result of intentions, proper function cannot constitute the condition of existence and persistence, and the theory needs to appeal to other conditions.[10]

A way out of this difficulty could be to waive the requirement of attributing a proper function to new artifacts and explore a functional conception of artifactual

[10] Etiological accounts could have at hand a possible way out of this tension. They can move towards linking the acquisition of proper functions to agents' intentions. This move could help them distinguishing between accidental and proper function and to use this distinction for ascribing functions to malfunctioning and genuinely new artifacts. However, as Vermaas and Houkes (2003, p. 282) remark, this move does not seem to help them at the end: "[a]scribing a function to an artefact just because it is the product of a desire that has the proper function to get itself fulfilled using the artefact seems all too easy: it provides no way of deriving conditions on the physical structure of the artefact which justify that it can perform the ascribed function." In the end, etiological accounts seem to be paying a high price for accounting for the "novelty aspect" in cases of genuine new artifacts.

novelty that did not have proper functions as its only axis. The pluralistic approach of functions proposed by Beth Preston (1998, 2003) follows this general strategy. According to her, we can attribute two sorts of functions to artifacts: proper and system functions (or systemic functions). As we have said, the proper function of an artifact is fixed by appealing to the causal history that led to the artifact being there with the structure that it effectively has – a history in which the intentions of designers and users obviously play a role. The proper function of an artifact is the "what for" of that artifact. The acquisition of these functions is a process analogous to the process by which biological features obtain their functions: a reproductive process contingent on success. On the other hand, system functions refer to tasks that the artifacts are able to carry out in terms of their materiality or the arrangement of their components. These depend on the capacities or dispositions that an artifact has and which may become manifest under certain circumstances, for example, when the artifact is employed in an idiosyncratic manner. Artifacts are intentionally propagated due to the (perceived) success that proper functions manifest in their actual use. However, systemic functions are the axis (Preston 1998, p. 250) on which a proper function is acquired for the first time, substituting the existing proper function or adding a new one to it.

From the perspective of this pluralist functional approach, we can say that an artifact is genuinely new when it carries out new functions, i.e., performs functions that did not exist before. The direct answer to the creative requirement *CR* is that a relevant causal reproductive history cannot be specified for genuinely new artifacts, *but* this does not mean you cannot appeal to a process of temporal drift through which systemic functions are exploited to replace existing proper functions or to generate or add proper functions (Preston 1998, p. 250). This approach does not intend to explain novelty or creativity in the field of artifacts; on the contrary, it postulates a process of change in which different phases emerge against a background of certain continuities. As a consequence, there is no creative production of a genuinely new artifact but a transition from one artifactual state to another in which functional properties are added, lost, or replaced. This process of artifactual selection appeals to the notion of systemic functions against a background of proper functions (i.e., once systemic functions become selected and fixed, they become proper functions). This position tells a story of intentional selection in cultural contexts in which *exaptations*,[11] products of intentionally selected systemic functions, acquire different degrees of fixity until the artifact is reproduced resulting in a new proper function. In the context of our argument, the most important conclusion about this approach can be expressed as follows: *CR* does not apply to this model of change through *exaptations*. On the contrary, the thesis that novelty is the result of an intentional selection process that operates on systemic functions goes against some of the assumptions that motivate *CR*. We could say that, in fact, this requirement

[11] An exaptation refers to a trait that evolves for a different function than its actual one or for none in particular. An exaptation occurs when an artifact's systemic function becomes an artifact's proper function.

and its demands are not real concerns for this theory. Moreover, it faces difficulties for understanding it as a constraint on any theory about the nature of artifacts.

Now, suppose that we accept an etiological theory that ascribes proper functions to items that have been *intended* by a designer to be used for a certain purpose. Is the mere fact that the designer's intentions contribute to the institution of the proper artifactual function sufficient to meet *CR*? Vermaas and Houkes (2003) think that this theoretical position can straightforwardly provide an explanation of it. The reason is that there would not be any difference between a *new* artifact and a standard one: we could ascribe to both cases a new proper function based on the designer's intentions. Suppose, further, that the proper function determines *kind* membership. It would follow that the intentions establish the identity criteria of the new artifact. The proper function is the *intended* function, determined, at least in part, by the intentions of its makers.

However, if the novelty of a proper function is explained by an appeal to intentions, what prevents the intentions of *users* from determining the new functions? Nothing seems to prevent it. Preston (2003) has argued that there is not any phenomenological grounding for distinguishing between these two types of intentions. Her argument shows on the basis of empirical examples that the designer's intentions do not differ in relevant aspects from the user's intentions.[12] The corollary of this discussion is that if the intentions of the designers are sufficient to assign a function to new artifacts (and to determine kind membership), so are the intentions of users. Thus, if a new use could be sufficient to attribute (or confer) a new proper function, then the distinction between proper functions and accidental functions would lose its meaning or collapse.

Preston (2003), in contrast to Vermaas and Houkes (2003), prefers to retain the normative and permanent character of proper functions and to introduce, as we have seen above, a new type of function to characterize artifacts: nonnormative, temporal systemic functions that account for the *nature* of new artifacts. However, this strategy can be taken as abandoning the novelty condition, since there would be no genuinely new artifacts; rather an artifact's novelty is explained as the appearance of a systemic function that becomes stabilized as a new proper (normative) function after a period of time. She seems to challenge the demand of the principle of novelty (and this would extend to our version of the creation requirement *CR*) because she believes that a principle like this depends on an individualist misconception according to which the determination of a proper function, as *intended function*, may be in the hands of individual agents. Anyone who would accept that the novelty requirement should be explained would commit herself to "a thoroughly individualist theory

[12] Preston's argument (2003, pp. 606–608) is articulated around the following issues: (a) designer's intentions do not embody a different cognitive structure than user's intentions; (b) user's intentions show the same degrees of creativity as the designer's ones; (c) user's intentions also involve the intentional transformation of the artifact's materials for performing a function; and (d) both the average designers' intentions and the average users' intentions do not mark the distinction between proper and accidental functions. For a detailed discussion of what she calls "the case of the recalcitrant prototype," see Preston (2006).

of function in virtue of its insistence that the intentional states of individual agents are sufficient by themselves for the establishment of the proper function of artifacts" (Preston 2003, p. 609). Preston claims that artifacts are cultural products subject to processes of reproduction and use in a social context. Thus, she concludes that the distinguishing feature between a proper function and an accidental one does not lie in the intentional aspects as such but in the individualistic intentional aspects. But, regardless of whether you insist (correctly, from our point of view) that proper functions are standardized functions that involve long social processes of reproduction and use, where the designers are not the only type of agents that contribute to the production of the artifact, the essential point is that new artifacts can only be properly characterized in terms of systemic functions, even if they are accidental, and these artifacts *could* indeed be the result of the *individual* intentions of designers and users. For anyone who accepts *CR*, the conclusion is that *CR* forces to adopt different theories for standard artifacts and new artifacts. Of course, there remains the alternative of challenging the relevance of *CR* for any theory on the nature of artifacts. In any case, this theorist has to deliver a plausible philosophical account that makes room for novelty and creation in artifactual realms.[13]

However, the problem of intentionalist conceptions of proper functions is more general. *CR* demands an account of both genuine novelties and the inauguration of new kinds and postulates that the determination of a proper function serves to define the conditions of identity and kind membership. So, if the proper function is fixed in part by the intentions of the designers (i.e., a proper function is an intended function), then we must explain how these intentions intervene in the determination of the boundaries of the kind. Of course, this does not mean that there are no other requirements, such as that the intended function be physically realizable and that there is knowledge available of why a certain physical structure could carry out a certain function (Vermaas and Houkes 2003, p. 287); what does seem clear is that what sets the normative criteria for the identity of the artifact (the conditions of being such-and-such an artifact and not another because it fulfills such-and-such a function) is the intentional dependency (that selects this functional structure and not another). Does this amount to a metaphysical dependence of the artifact, as characterized by its intended function, on the intentions of the designers/makers? The theory of proper functions as intended functions seems to share with intentional

[13] Even if the details of Preston's theory cause trouble for understanding novel artifacts, there are other theories that one could imagine to face this challenge. One of them, for example, would just be that functions are systemic functions. On this view, the functions of a novel artifact could be understood along the following line: the function of x is what x does (Kitcher 1993); therefore, there is a new function that is performed, where function equals something that is done. Another theory would be disjunctive, i.e., functions are either proper functions (etiological) or systemic functions (non-etiological). Of course, this last type of theory would have to adopt different theories for standard artifacts and newly created artifacts. However, a demand for uniformity may not be an issue here, given that there seems to be a relevant difference in view: the function of a well-know artifact is historical in a way that the function of a novel artifact isn't. Nonetheless, these two theories would make it impossible for us to account for the normative dimension of the function of the new artifact. (Thanks to an anonymous referee for pressing this point.)

theories of artifacts these crucial aspects with respect to the explanation of what makes an artifact what it is, in particular in the case of a newly created one. As noted by intentional-historical theorists, the idea that artifacts are characterized by their intended function is just a specification of a more general fact that the makers are the ones that determine kind membership (regardless of the fact that sometimes they specify an intended function, which is not performed by all artifacts) (Thomasson 2007, p. 58).

So far it seems as if functional-intentionalist and historical-intentional theories of artifacts could accommodate *CR*. However, the next section will argue that this is not the case.

7.4 Intentions, Creation, and the Nature of Artifacts

In the two previous sections, we have argued that any conception of the nature of artifacts and artifactual kinds must be able to meet the challenge of characterizing the ontological nature of newly created artifacts. It was also suggested that the identity of proper function does not seem to be a good criterion to characterize them ontologically. It becomes very difficult to attribute a proper function to new artifacts, the first members of a new artifactual kind, because this attribution must depend on a causal and reproductive history. Two routes seem to be open at this point. One of them might be to apply a different notion of function to prototypes and new artifacts – e.g., the notion of systemic function. However, one may wonder why, if this notion is enough to pick out a new artifact, it is not enough to identify what artifacts are in general – i.e., objects ontologically characterized by the fact that they perform a certain systemic function. A second route might be to recognize the relevance of the intentions of designers and/or users in identifying the function that picks out members of a kind. In this case, if the artifacts are individuated in terms of function, they are so according to their intended function. What confers unity to the members of an artifactual kind? The answer is: the intention under which they have been produced. And this would be so for the *first member* of the kind. However, once one has accepted that the proper function of an artifact is its intended function we have taken a first step away from a generalized functional theory towards a clearly intentional one. It would suffice for giving up a purely functional theory as a general analysis of the nature of artifactual kinds to realize that some artifacts (such as works of art) are not fundamentally characterized in terms of their function (intended or not) and that members of certain functional kinds have not even been intended as having the function that identifies them. Something can be a pencil even if its maker wants that it never be used as such. As Thomasson remarks having these cases in mind, "sharing an intended function cannot be essential to the nature of an artifactual kind" (Thomasson 2003, p. 594). The fact that some artifacts are well characterized by the *intended* function is just a consequence of a more general theory that could accommodate nonfunctional artifacts and those members of the kind that are not intended as such. This theory could take the shape of the so-called intentional-historical theory of artifacts, a theory that definitively gives to the creator's intentions the key to determine the identity of the members of an artifactual kind. In this section, we will examine how this theory faces the challenge of novel artifacts.

Proponents of the intentional-historical account take as a starting point the "production" of particular artifacts belonging to an existing kind, known by its producer. Bloom (1996, p. 10) suggested that the categorization of an artifact as a member of a certain kind can be explained by the fact that this artifact is a result of the same type of intention that led to the production of other members of its kind. However, he talks of *creating* something of the *same* kind as current and previous members. It is obvious that this idea does not address the question of the creation of artifacts insofar as they inaugurate a kind, since this one preexists, in a sense, each new instance of it. So, the intentions that intervene in the production of the kind of artifact in question, and in determining its identity as such artifact, can be here not only relevant but also determinant. The intention, whose content involves a certain artifactual concept *A*, controls the production of an *A* and also determines the kind of object it will be, i.e., an *A*, to the extent that successful realization leads to an object that satisfies those characteristics that make it to belong to the kind of *As*. We will expand this further.

The main thesis of this approach may be stated in the following way: the creation of a particular artifact is the outcome of the realization of the agent's intention, whose content involves an idea of the kind of thing that is being created. We can clarify this thesis as follows:

Given a maker *M* and an artifact *A*, *M* creates *A* only if the following conditions are satisfied:

1. M has the intention *I* of creating *A*.
2. The content of *M*'s intention *I* to create *A* involves an idea of what kind of thing an *A* is.
3. *M* carries out intention *I*.
4. The intention *I* is successfully satisfied.
5. *M* is able to assess the degree of success of his intention *I* to create *A*.[14]

What do we mean when we say that the intention of the creator of an artifact should involve an idea of the artifact in question? It is said that the creator must have a concept or "some description" (Hilpinen 1993, p. 157) of the artifact she will produce. This concept contains the set of properties (whether functional or material or others) that characterize the artifact and that will be imprinted on matter. The core of this concept is given by a description that relates to the kind of thing that is going to be created. Thus, if a maker intends to create a corkscrew, the contents of her intention will not only collect, for example, the following properties: "being able to remove corks," "being of a white color," "being a material resistant to tampering,"

[14] This fifth condition captures our ordinary intuition about how to distinguish between someone who intentionally produces an artifact and someone who merely gets some result by toying around without any previous idea about what she is trying to make.

and "being light," but will also contain a description such as "that it belongs to the kind or type of corkscrews." Obviously, a description of the kind involves properties that are substantive to the specification of what the artifact effectively is.[15] This description helps to determine the identity of the created object, to distinguish it from objects of other kinds, and especially to assess the degree of success with which the intention has been realized; in fact, that the creator successfully produces a corkscrew depends on whether the object produced satisfies the idea of the kind of thing that the creator was trying to produce and on whether she gives her acceptance that this is indeed the case. Only if the creator *possesses the concept of the artifact* to be produced is she in a position to impose on matter a set of properties and to evaluate and accept some degree of success in that intentional realization. If this does not happen, the creator would find herself blind regarding her productive activity, i.e., acting without any kind of mental direction since she would not be guided by any notion of the kind of object she is going to create. Likewise, she could not distinguish her achievements from other results of her productive activity or interpret them. Since, if the creator does not possess the concept corresponding to the artifact she intends to produce, she is not able to verify and evaluate whether the properties that the concept fixes as essentially relevant for being such an artifact have been given expression in matter as actual achievements of her intentional activity.

This constitutive dependence of an artifact on the content of the intention of its creator has been recently formulated in terms of necessary conditions by Thomasson (2007). According to her:

> Necessarily, for all x and all artifactual kinds K, x is a K *only if* x is the product of a largely successful intention that (Kx), where one intends (Kx) *only if* one has a substantive concept of the nature of Ks that *largely* matches that *of some group* of prior makers of Ks (if there are any) and intends to realize that concept by imposing K-relevant features on the object. (Thomasson 2003, p. 600; emphasis in the original)

As can be quickly perceived, Thomasson's formulation suggests that the concept of the kind of thing to be created, which is an essential part of the contents of the creator's intention, should be a shared concept, namely, it must adjust to a large extent to the concepts that previous makers of that kind of thing possess. And while the formulation makes clear that this must be the case if there are any previous substantive concepts of that kind of thing, the intentional-historical approach does not seem to explore how to understand the creation of artifacts that do not fall directly under already existing kinds of artifacts. On the contrary, it is an approach that at a first sight is fully adequate to deal with cases of production of artifacts that belong to already existing kinds.

Because of that, the intentional-historical approach tends to raise the issue of the creation of new kinds of artifacts only as a modified form of production of an artifact belonging to an already existing kind. In fact, Thomasson acknowledges as

[15] This intention cannot be understood as a transparent one. To employ Thomasson's words: "as a bald intention to make 'one of these' (pointing to a sample)" (2007, p. 58).

a challenge to her formulation of the theory the question of what would happen if the intention of the maker differs somewhat from the intentions of previous makers and ends up reflecting (i.e., the intention) an idiosyncratic concept of the maker, therefore realizing an artifact significantly different from its ancestors. The proponents of this approach only suggest that a substantive concept of the nature of an artifactual kind can gradually change over time and can do so in more than one dimension. However, this does not seem to be an explicit explanation of what we have called the novelty condition and creation requirement.

The intentional-historical theorist approaches the creation of artifacts and artifactual kinds *from above*. The theory appeals to well-defined intentions, the content of which is articulated around the concept of the kind of object to be made. Then, it adds a practical condition: the content of the intention must be successfully realized. The intention essentially involves a substantive concept of what an *A* is and also some understanding of the kind of properties that are relevant to belong to the kind *A*. However, in the case of genuine creation, when outlining the new kind, there is nothing right or wrong about what an *A* may be or what the maker can appeal to in order to guide and evaluate her activity (Thomasson 2007, p. 60). What aspects are relevant, then, for being an *A*? What aspects does the maker intend to impose on the object? They can be no more than those aspects *stipulated* by the maker as relevant to identify the kind of artifact that is to be created. The conditions for normative success are fully established by a stipulative act on the part of the maker: "Thus she creates not only an artifact, but delineates a new artifactual *kind*, complete with normative success conditions for creating something of that kind" (Thomasson 2007, p. 60).

Of course, what is not at stake here is the fact that the very *existence* of an artifact, inasmuch as it belongs to this new kind, depends solely on the mental activity of the maker, as if a sort of "conjuring trick" (Thomasson 2009, p. 195) would have given life to the artifact. Other factors – such as materials or maker's abilities – are essential in bringing about an artifact. As we have seen, an artifact that inaugurates a new kind, or which cannot be identified as belonging to already established lineages, is *what it is* given the conditions of individuation determined by its kind. And it seems to be in the hands of the designer/maker, according to the intentional-historical conception, to define the kind by stipulation and, therefore, to define also the relevant conditions of individuation. Would that be enough in order to characterize the ontological nature of a new artifactual kind?

Let us suppose that it is sufficient to stipulate which aspects determine the kind. In this case, the success or failure of the execution and, therefore, the effective production of the artifact do not affect how the maker's intention and the concept of an *A* that is involved in the intention are determined. At the same time, the maker seems to know what it is to be an *A*. It is purely contingent that the maker manages (or not) to succeed in carrying out the intention. It could be that through a strange mechanism, the object appeared in the world with those traits that substantially characterize an *A* according to the concept that guides the intention of maker *M*, i.e., that they correspond to what has been stipulated as being an *A*, and that was not a result of the realization of the intention of the maker. Is, in this case, *A* defined merely in terms of those aspects relevant to be an *A* that the maker *M* has stipulated? Moreover, the failure in the execution of the concept of an *A* would also not undermine the fact that the kind is already delineated in the stipulation. If so, then it seems that the possession of a concept by the creator would be sufficient to delineate the kind, whether or not there is an *A*.

At this point, there are two additional problems. On the one hand, it is odd to say that the kind is fully determined prior to there being a member of it. This would seem to entail that there could be kinds whose extension is empty and has always been empty. This remark is not about the possibility of having a concept with an empty extension. We all know that there are many; the question concerns the possibility of *sortal* concepts grasped with independence of the existence of any member of the kind. Biological kinds, if they exist, could provide a good analogy at this point. We can have sortal concepts that identify members of extinct biological kinds (let us say, species), in a way that enables us to say that these sortal concepts have empty extensions. But in this case, those who grasp the concept have an idea of what it is to be a member of the kind. The question with the concepts of nonexisting artifacts is that our grasp of them does not give us a sufficient criterion to determine which sort of object would fall under its extension.

One option is to think that concepts of artifacts were already there, in something like a Platonic heaven of universals which we identify with the kinds and that we "intuit" through our stipulations, regardless of existing artifacts. Even if we accept the Platonic ontology of universals and even if we apply it to artifacts (as Plato himself seems to have done), it is very doubtful that our stipulative practices get to *grasp* the relevant identity conditions of the artifact in question, which are given by its Platonic essence. Another possibility is to argue that concepts as such are created in advance by the inventor (before making a particular artifact and probably before *being able* of making it). The creation of a concept will suffice to delineate the identity conditions of a created abstract object (a sort of abstract artifact). Real concrete artifacts, once produced, would share the very same identity conditions of the "abstract" members of the artifactual kind so created. But a theorist that would argue along these lines would yet need to give us some idea about how mere stipulations are sufficient to create an abstract kind of artifacts that will coincide with the kind as concretely real.[16]

On the other hand, it is doubtful that the mere fact of stipulating success conditions on the basis of what is relevant for being an *A* may, in turn, be sufficient for the genuine possession of an artifact's concept. Did Leonardo perhaps have a concept of something like a "plane" when he drew his sketches of flying machines? Were his *ideas* a way of delineating an artifactual kind by stipulating features

[16] We thank an anonymous referee for suggesting to us this option. It is clear that inventors can have more or less clear concepts about what they are trying to bring about, but we think that it is also true that it is difficult to see in what sense these concepts, insofar as they just gather those features stipulated by the creator, delineate the very existence of an artifactual kind (abstract or concrete). Obviously, they imagine a possible artifact, but this raises different (and also interesting) ontological problems.

relevant to *A*? Here, intuitions could be in conflict. We claim that there is a clear-cut distinction between mere ideas of possible artifacts and those sortal concepts of artifacts that are needed to delineate a kind. A mere idea of an artifactual possibility, even if it is well defined and could itself be, in a sense, true of the artifact[17] (i.e., it would lead to an efficient realization if all technical and skill conditions were in place), does not constitute by itself the grasping of the sortal concept of an artifact or, at least, is not sufficient for possessing a concept of an artifact that will effectively and appropriately guide its production. Besides, sketches and suggestions on possible new artifacts, whose conditions of realization are not yet well grasped, are easy to get and that would populate our world with artifact concepts that would identify artifactual kinds with empty extensions.[18]

Now suppose that stipulation is not sufficient to identify the relevant features that the maker intends to impose on objects and that we may require some other condition related to the fact that the intention is effectively executed and the artifact comes into being with a form, function, or any other distinctive features proper to the object. If so, then there are dimensions of what it is to be, for example, an *A*, that are not constituted by the substantive idea that guides the production of this artifact. Let us be clear, this does not deny that there is some conceptual dependence constitutive of the artifact and the metaphysical nature of the kind in relation to the intentional states of the makers (and users). Nonetheless, it is not enough to appeal to a substantive prior idea as a guide to production in the case of the creation of artifacts. On the other hand, there are aspects of the metaphysical nature of the kind that are not exhausted by the intentions and concepts of the makers. Briefly, its nature cannot be delineated just in terms of the features characteristic of the kind imposed on matter by the designer/maker, because some of the ontologically decisive features come to light through the effective practical process of producing a particular of the kind that meets the success conditions of the intention.

The core of the difficulties of the intentional-historical conception in relation to *CR* is that it does not seem easy to attribute to the maker the possession of a genuine *concept* of the artifact that will be created. On what basis does *M* have the concept of an *A* such that it can, on the one hand, effectively guide the production of members of the kind and, on the other, define the nature of the kind? Does the producer understand what makes an artifact a possible member of the kind as it is identified by the concept that guides its production? Does the realization of an artifact with stable identity conditions that are, so to say, "discovered in bringing previous intentions to effective realization" not add anything to the conditions of possession of the

[17] Probably, if it is true of an artifact in this sense, it is just by chance.

[18] An interesting question is about whether and how one could delineate a space of possibilities of artifacts that resemble a logical space of concepts. It is true that part of our work as designers and makers is to explore a space of technical possibilities that probably grows from our understanding – that is of the concepts – of the artifacts already existing in our cultural world. But this does not say yet anything about the plausibility of the notion of possessing a sortal concept of an artifactual kind merely through the grasping of a certain possibility already given in this space of artifactual possibilities.

concept? And does the process of effective appearance of the artifact not add anything to those very conditions of possession? In short, intentional-historical theories that regard the carrying out of the intention as a mere process of execution are irrelevant when it comes to determining the conditions for success of the very activity of creation and production. The execution can be successful or not, but it does not teach us anything about what the object is and whether it is consistent with the concept of an *A*. It lacks ontological import.

We could get out of this difficulty by claiming that, in the case of the creation of artifacts and new artifactual kinds, there is still no concept of the kind that will be created; on the contrary, the process itself is a process of learning that combines discovery and invention, in which the stabilization of an effective artifact allows the constitution of the kind and secures the conditions of possession of the concept that describes the kind. Only then, once you actually identify a particular member of the kind and fix the conditions for possession of the concept, would it make sense to adopt a particular intention to produce an *A*.

There is another way of formulating the problems that face any intentional conception of the nature of newly created artifacts. The knowledge the designer/maker has of what it takes to be an artifact *A* is dependent on the fact that the designer/maker stipulate what it is "by establishing success criteria for her activities" (Thomasson 2007, p. 61). But this does not mean that the designer/maker already knows which the normative conditions of success of that very kind of artifact are. In other words, we cannot apply to the maker's knowledge of artifacts a model that considers it as a sort of knowledge *by fiat*. Knowledge of the normative conditions of success cannot be established by stipulation, so we need some knowledge obtained by observational means or by other more substantive means. Making provides by itself knowledge of what is made, because of the kind of understanding that the maker exhibits and acquires in the very activity of making the artifact. In a sense, our knowledge of what we are making is also dependent on the interaction with the object itself, with other objects within our material culture and on our control of causal patterns in the world. The success conditions that delineate the nature of the kind that is created are given to the creator in these circumstances.

Thus the historical-intentional view faces a new dilemma: on the one hand, if it wants the concepts to be effective in the process of bringing about of an artifact, the view is committed to defending that those concepts involved in the intentions of the makers of artifacts are rich enough; on the other hand, the historical-intentional view has to acknowledge that the maker only has at her disposal a concept whose determinations are the result of a stipulative act by the maker. In a sense, it is very easy for a maker to possess a certain concept of an artifact, because it seems to depend only on a sort of understanding tied to her stipulative acts. It just includes some normative requirements the object should meet. On the other hand, if this would not be enough to possess the concept of the artifact that is being created, then new features relevant for the artifact to belong to the kind are needed beyond the mere conceptual dependence on the maker's intentions. Surely, the practical process of bringing about a new particular with the features that actually has at the end of the process contributes to delineate the individuation conditions of the members of the kind, because it is in this process that the maker acquires the understanding of the success conditions characteristic of such kind of artifacts. This means that it is only at this moment that she grasps the concept of the artifact and the new kind is delineated. In other words, contrary to the claims of the intentional theorist, *CR* is not easily satisfied by just appealing to a conceptual dependence of newly created artifactual kinds on the conceptually informed intentions of the designers/makers.

7.5 Conclusion

We have argued in this chapter that two types of theory on the nature of artifacts, i.e., functional and intentional theories, do not appropriately satisfy to the *creation requirement*, which demands an account of the appearance of genuinely new artifacts resulting from intentional creative processes. If we choose to account for the nature of artifacts in terms of proper functions, the resulting explanation excludes an account of the nature of newly created artifacts. If we claim that the emergence of a new artifact depends on the possession of a new concept of that artifact, then we face a dilemma: either we need to excessively weaken the conditions for possessing a concept of an artifact or we need to concede that the constitution of the newly created kind cannot depend completely on the concept. We have argued that the reason for this dilemma is fairly simple: in the case of the creation of artifacts, the creators cannot possess the concept that will guide the production of the artifacts.

As a way of concluding, we would like to point out that, although it was not our goal to raise questions about creative processes, we believe that, at least regarding some of the artifacts that populate our material culture, it is necessary to account for the fact that what they really are is not something separable from the fact that they are the result of intentional creation. At least four aspects characterize a situation in which the emergence of creative novelties is possible. And all these aspects have to be taken into account to satisfactorily deal with the ontology of artifacts:

1. Creative production sets out from the idea of what we want to achieve; it starts from certain requirements that must be met and the satisfaction of which cannot be attributable to chance.
2. Creative production is not guided, nonetheless, by a concept that describes the traits that determine the character of what is being produced, since genuine creation institutes, so to speak, a hitherto nonexistent rule (it does not fit standards or concepts previously held by intentional agents).
3. Creative production is done under certain conditions of ignorance in relation to the created product.
4. Creative production is not carried out under conditions of complete blindness, as intentional agents have skills, knowledge, and previous solutions (present in a concrete social and cultural setting) that may suggest ways to proceed and solve problems.

Creating an artifact involves learning about what intentions are within our reach and what goals are achievable. Our main insight can be summarized as follows: the sort of understanding that is linked to the possession of a concept of an artifact (and not a mere idea of a possible and imagined artifactual object) is acquired in the actual effective production through which it is "discovered" under what conditions the artifact adequately performs what it is *supposed* to do. One might ask whether this is not circular. How can we set what the artifact is "supposed" to do without resorting to the history of its ancestors (functional-reproductive conception) or to a stipulation by its "creators"? This is what makes it difficult to satisfy the creation requirement. The reason is that you are forced to connect two aspects involved in the sort of normativity we usually recognize to be at work in the field of artifacts: on the one hand, the fact that, somehow, there is still not a norm (relating to what it is to be an *A*) to explain what the success of our production consist in; on the other hand, the fact that the subject cannot be in a position to understand what success really is before a design becomes a reality. The creator, in a sense, imposes the norm, but what the norm dictates is not intelligible without effective and real success. As we have seen, the intentional-historical account emphasizes the first aspect. The functional-reproductive conception emphasizes the second one. Like other philosophical challenges, striking a balance between both dimensions may demand a new way of understanding the ontology of artifacts.

Acknowledgments A previous version of this chapter was presented at a seminar on the metaphysics of artifactual kinds, organized at the Delft University of Technology. Comments and questions have helped us to sharpen and clarify our ideas and arguments. Fernando Broncano has read previous versions of this chapter. We thank him for his valuable and insightful suggestions. We also thank the editors for searching criticisms. This research was funded by the Ministerio de Ciencia e Innovación (Spain) (FFI2009-12054). It also received partly financial support by CONICET (Argentina).

References

Baker, L. R. (2007). *The metaphysics of everyday life*. Cambridge: Cambridge University Press.

Bloom, P. (1996). Intention, history, and artifact concepts. *Cognition, 60*, 1–29.

Briskman, L. (1980). Creative product and creative process in science and art. *Inquiry, 23*(1), 83–106.

Grandy, R. (2007). Artifacts: Parts and principles. In E. Margolis & S. Laurence (Eds.), *Creations of the mind* (pp. 18–32). Oxford: Oxford University Press.

Hilpinen, R. (1993). Authors and artifacts. *Proceedings of the Aristotelian Society, 93*, 155–178.

Kitcher, P. (1993). Function and design. *Midwest Studies in Philosophy, 18*, 379–397.

Preston, B. (1998). Why is a wing like a spoon? A pluralist theory of function. *The Journal of Philosophy, 95*, 215–254.

Preston, B. (2003). Of Marigold beer: A reply to Vermaas and Houkes. *The British Journal for the Philosophy of Science, 54*, 601–612.

Preston, B. (2006). The case of the recalcitrant propotype. In A. Costall & O. Dreier (Eds.), *Doing things with things. The design and use of everyday objects* (pp. 15–27). Aldershot: Ashgate.

Soavi, M. (2009). Realism and artifact kinds. In U. Krohs & P. Kroes (Eds.), *Functions in biological and artificial worlds: Comparative philosophical perspectives* (Vienna series in theoretical biology, pp. 185–202). Cambridge, MA: MIT Press.

Thomasson, A. L. (2003). Realism and human kinds. *Philosophy and Phenomenological Research, 67*, 580–609.

Thomasson, A. L. (2007). Artifacts and human concepts. In E. Margolis & S. Laurence (Eds.), *Creations of the mind* (pp. 52–73). Oxford: Oxford University Press.

Thomasson, A. L. (2009). Artifacts in metaphysics. In A. Meijers (Ed.), *Philosophy of technology and engineering sciences* (pp. 191–210). Amsterdam: Elsevier.

Vermaas, P., & Houkes, W. (2003). Ascribing functions to technical artefacts: A challenge to etiological accounts of functions. *The British Journal for the Philosophy of Science, 54*, 261–289.

Chapter 8
Metaphysical and Epistemological Approaches to Developing a Theory of Artifact Kinds

Thomas A.C. Reydon

Abstract In this chapter, I examine whether the philosophy of natural kinds can yield useful clues for the development of a theory of the metaphysics and epistemology of artifact kinds. In passing, I explore the prospects for a unified account of kinds that applies to natural kinds as well as artifact kinds. As a generally accepted, satisfactory theory of natural kindhood is still lacking, both these prospects appear dim. I review parts of the recent history of philosophical work on natural kinds, particularly in philosophy of biology, and highlight a development that I call an "epistemological turn" in thinking about kinds. I argue that recent work following this epistemological turn shows promise when it comes to the development of a theory of artifact kinds but that there still are problems specific for the artifactual realm which theories of artifact kinds must address.

Keywords Artificial kinds • Artificial kind/Artifact kind distinction • Boyd, Richard • Epistemological turn • Essence • Engineering disciplines • Homeostatic Property Cluster Theory (HPC-theory) • Kind theories, metaphysics-oriented • Kind theories, epistemology-oriented • Locke, John • Mind-dependence • Natural kinds • Natural kindhood, criteria for • Species

8.1 Introduction

What is the nature of artifact kinds? What determines the kind identity of individual artifacts and the boundaries of artifact kinds? Which epistemic roles do references to artifact kinds perform in reasoning and knowledge production, and what enables us to use them in these roles? Although the topic of artifact kinds is increasingly

T.A.C. Reydon (✉)
Institut für Philosophie, Leibniz Universität Hannover,
Im Moore 21, 30167 Hannover, Germany
e-mail: reydon@ww.uni-hannover.de

M. Franssen et al. (eds.), *Artefact Kinds: Ontology and the Human-Made World*,
Synthese Library 365, DOI 10.1007/978-3-319-00801-1_8,

receiving attention from philosophers, in particular in the domains of analytic metaphysics and the philosophy of science and technology, we are still far from a consensus on how questions like the aforementioned should be answered.

One way to address such questions about artifact kinds is to look for clues in the available literature on parallel questions that have been posed with respect to kinds in the natural domain. Philosophers have long been concerned with the metaphysics and epistemology of natural kinds in the various contexts in which these feature. Thus, one might hope that an account of natural kinds and related issues – such as classification and generalization in the sciences – is readily available for application in the artifactual domain. Perhaps even a unified account of kinds that applies to natural as well as artifact kinds could be achieved. At the very least, one would expect that the work philosophers have done on the topic of natural kinds can provide clues for the development of a theory of artifact kinds.

As a generally accepted, satisfactory theory of natural kindhood is still lacking, however, both the prospects for finding a readily applicable theory of kinds in the literature and for obtaining a unified theory of natural and artifactual kinds appear dim. In the present chapter, I shall explore to what extent such hopes and expectations are warranted. My aim is to draw lessons from the philosophy of natural kinds for the project of developing a metaphysical and epistemological theory of artifact kinds. Conversely, there probably also are lessons to be drawn from the case of artifact kinds for the ongoing natural kinds debate – but that endeavor will largely have to be left for future work.

I shall begin in Sect. 8.2 by examining the philosophical tradition of thinking about natural kinds in which metaphysical criteria for characterizing natural kinds are being sought. This search seems inspired by a commonly accepted dichotomy between natural and artificial kinds that in fact, I shall argue, has led the discussion on kinds in a wrong direction. In Sect. 8.3, I shall review some recent work on the topic of natural kinds – in particular from the philosophy of biology – and highlight a development that I call an "epistemological turn" in thinking about kinds. Due to problems that metaphysics-oriented approaches to developing a theory of natural kinds are confronted with, philosophers of science are increasingly approaching the issue by searching for criteria that characterize natural kinds from an epistemological point of view, instead of looking for the metaphysical criteria that distinguish natural kinds from other kinds of kinds. I want to suggest that this epistemological turn marks a more promising direction in the philosophy of kinds and classification, including artifact kinds. Here, the issue is approached by first elaborating epistemological criteria for natural kindhood and kind membership and then constructing a metaphysics that fits this epistemology best. In Sect. 8.4, I shall examine how the most prominent contemporary representative of this epistemology-oriented approach, Richard Boyd's Homeostatic Property Cluster theory, would work out if applied to artifact kinds. In so doing, I shall point to some problems that arise specifically for kinds in the artifactual realm and will have to be resolved before such an epistemology-oriented approach can be successful. I shall close in Sect. 8.5 by highlighting some advantages of taking an epistemology-oriented approach to elaborating the metaphysics and epistemology of artifact kinds over taking a metaphysics-oriented approach.

8.2 Artifact Kinds: Not Natural, but Not Artificial Either

The philosophical concern with the nature of kinds and classification largely stems from two sources. First, there is the long-standing metaphysical quest to provide an account of what kinds of things there are in the world we live in – that is, what the "furniture of the world" (Bunge 1977) is made up from. Second, there is the epistemological interest in how our classifications of things, phenomena, events, etc. relate to the world "out there" and whether some ways of classifying better agree with the facts than others. These issues have been central questions of philosophy since its early beginnings (Plato and Aristotle have already addressed them), but with the rise of philosophy of science as a philosophical specialization, they have come to lead a life of their own in what is sometimes called the "philosophy of classification" (Ereshefsky 2001).

In both metaphysical and epistemological investigations of kinds and classifications, the contrast between on the one hand natural kinds and natural systems of classification and on the other hand artificial (or conventional or nominal) kinds and artificial systems of classification traditionally plays an important role. Here, natural systems of classification are thought to represent aspects of the actual state of affairs in nature, in contrast to artificial classifications that merely are groupings that we humans use because they suit our purposes.[1] This dichotomy has been an element of the philosopher's toolkit at least since the appearance of Locke's *Essay Concerning Human Understanding*, where the two kinds of kinds are contrasted with respect to their mind-(in)dependence.

In the *Essay*, Locke famously introduced the distinction between the real essences of things ("the very being of any thing, whereby it is, what it is") and their nominal essences ("that abstract *Idea*, which the General, or *Sortal* [...] Name stands for").[2] A thing's real essence is something about the thing in itself that makes it to what it is, independently of the existence of knowing subjects that identify the thing as a particular individual or as something of a particular kind. In contrast, nominal essences are ideas or concepts in the minds of knowing subjects who use them to group things in ways that happen to be epistemically or practically useful to them. As Locke explains, knowing subjects use concepts to

> enable themselves to consider Things, and discourse of them, as it were in bundles, for the easier and readier improvement, and communication of their Knowledge, which would advance but slowly, were their Words and Thoughts confined only to Particulars.[3]

---

[1] Cf. the quest for a "Natural System" in biology that played a role in the *Origin of Species* (Darwin 1859, p. 413). For a historical account of the concepts of "natural kind" and "natural system," see McOuat (2009).

[2] Locke, *Essay*, Book III, chap. III, §15.

[3] Locke, *Essay*, Book III, chap. III, §20.

For Locke, nominal essences played a much more important role in our classificatory practices than real essences, as on Locke's view the human cognitive faculties did not allow us to know the real essences of things.[4]

On Locke's dichotomy, any group of things is either a mind-independent natural kind or a mind-dependent artificial kind. Here, the notion of mind-independence refers to the factor that determines the kind-membership of an entity: for natural kinds, this is an intrinsic property of their member entities (i.e., a property these entities have independently of cognitive subjects), while for nominal kinds, this is a mind-dependent idea on the basis of which we group things together. Things that share the same real essence independently of human cognition objectively (that is, mind-independently) can be said to belong together because it is "Nature in the Production of Things" that has made them alike, independently of any classificatory activity of knowing subjects.[5] In the centuries following Locke's *Essay*, such kinds have often been identified with natural kinds as being precisely those kinds of things that exist "out there" in the world, independently of human cognition, human actions, and human interests. Consider, for example, the second sentence in the *Stanford Encyclopedia of Philosophy* entry on natural kinds: "To say that a kind is *natural* is to say that it corresponds to a grouping or ordering that does not depend on humans" (Bird and Tobin 2010). Artificial (or conventional or nominal) kinds, in contrast, fully are products of the "Workmanship of the Understanding"[6]: they are fully dependent on humans and their epistemic interests, against the background of which people group things under the various concepts they employ. Following this dichotomy, the philosophy of natural kinds is primarily a question of metaphysics, not epistemology: it is about finding out which criteria objectively make things to the kinds of things they are.

Note that if this traditional dichotomy between mind-independent natural kinds and mind-dependent artificial kinds is correct, it is unlikely that among the available accounts of natural kinds there will be an account that also applies to artifact kinds. Artifacts do not exist "out there" in the world independently of human actions and interests but are mind-dependent objects: they are conceived, designed, and made

[4] Locke, *Essay*, Book III, chap. VI, §9; Book IV, chap. VI, §12. There exists controversy among Locke scholars on whether Locke actually embraced the view that there are real kinds in nature (e.g., Uzgalis 1988). While he might not have embraced that view, it did play an important role in his thinking about kinds. As such, his dichotomy between real and artificial kinds was more a starting point for his argument than a position he argued for in the *Essay*.

[5] Locke, *Essay*, Book III, chap. III, §13. Of course the identification of the real essences of things, the grouping of things with the same real essence into kinds, and the naming of kinds remain acts of knowing subjects. Nature just makes it so that multiple things have the same real essence. But it is in this sense – that a uniquely privileged basis for grouping things into kinds is given by nature – that a kind can be said to exist in nature and be mind independent.

[6] Locke, *Essay*, Book III, chap. III, §13. Thus, the principal difference with natural kinds is that here the basis for grouping things into kinds is not given by nature. There doesn't seem to be any privileged way of grouping things into kinds, so the ways in which things are grouped fully depends on human interests (or at least does so to a much higher degree than in the case of natural kinds).

by human beings who have particular purposes in mind (e.g., Verbeek and Vermaas 2009, p. 165; but see Sect. 8.4 for some nuances). By consequence, the same holds for artifact kinds: if artifacts are mind-dependent things, the ideas in the minds of designers, makers, and perhaps users are crucial factors determining the kind membership of artifacts (Franssen et al. 2009, Section 2.5). Both artifacts and artifact kinds are "creations of the mind" (Margolis and Laurence 2007) and, therefore, must be metaphysically distinct from natural kinds. Because "the very being of an artifact, whereby it is, what it is" depends on the minds of at least the artifact's designers and makers and is not a mind-independent Lockean real essence, there is no reason to expect theories of natural kinds to apply to artifact kinds.

However, rather than accepting this conclusion, I take the case of artifact kinds as suggesting that there is something wrong with the traditional dichotomy. Although artifact kinds do not seem to constitute a subgroup of the natural kind category, it isn't plausible either that artifact kinds belong into the artificial/nominal kind category. Surely artifact kinds are more than just groupings that, loosely quoting Locke, "enable humans to consider things in bundles for the easier and readier improvement and communication of their knowledge." Once a particular kind of artifact has been conceived and the first members of the kind have been designed and produced, such a kind *does* have something close to the objective existence that characterizes natural kinds. From the perspective of individual human beings most kinds of artifacts exist as real kinds "out there" in the world just as natural kinds do: new human beings are born into a world the furniture of which consists as much of hammers, cars, nation states, etc. as it consists of electrons, gold atoms, tigers, and the like. Accordingly, several philosophers have come to count artifact kinds together with natural kinds as real kinds (e.g., Boyd 1999a, b, 2000; Millikan 1999a, 2000; Elder 2007; Thomasson 2007; Carrara and Vermaas 2009).[7] While not being natural kinds, artifact kinds seem metaphysically less distinct from natural kinds than is often thought.

The case of artifact kinds thus fails to fit a central part of the traditional – and today still widely accepted – metaphysical framework for thinking about kinds. While this might be taken to imply that much of what I have said above about artifact kinds is wrong and artifact kinds actually *are* nothing more than convenience-based groupings of things, I suggest that a better implication to draw is to think of the traditional dichotomy between mind-independent natural kinds (where mind-independence means that nature provides us with a privileged basis for grouping things into kinds) and mind-dependent artificial kinds as misconceived. Thinking of artifact kinds as merely convenience-based groupings seems inadequate to how we actually classify artifacts in a world in which kinds of artifacts are as much part of the furniture of the world we find in place as are kinds of naturally occurring things.

---

[7] Boyd's examples include money and political systems (feudalism, parliamentary democracy, monarchy, etc.). Although these are not typically counted as artifacts, it is unclear why they should not count as such (see Sect. 8.4 below). Millikan mentions car models such as the 1969 Plymouth Valiant 100 as real kinds. Losonsky (1990) made a similar point, arguing that both natural objects and artifacts have natures that characterize them and can be investigated scientifically.

Thus, one of the traditional metaphysical assumptions about what it is to be a natural kind that has shaped much of the philosophical discussion on natural kinds – i.e., that to be a natural kind is to be a mind-independent kind, in contrast to kinds that do depend (much more strongly) on (human) minds – seems to have to be discarded.

For this reason I am pessimistic about the prospects for success of a search for metaphysical criteria for natural kindhood – as we now have reason to consider one of the central tenets of traditional metaphysical accounts of natural kindhood to be defective – and want to suggest that a better approach might be to address the epistemology of kinds first and then go on to construct a metaphysics that fits the epistemology.[8] Indeed, a reorientation that is in accord with my skepticism is visible in recent work on natural kinds in the philosophy of science, constituting a development that opens up new prospects for thinking about artifact kinds too.

8.3 An Epistemological Turn in Thinking About Kinds

A central part of the metaphysical tradition that was problematized above is the assumption that the notions of "natural kind" and "real essence" are inseparably connected. Natural kinds, on this view, are kinds of things that all share the same essence: *all* members of a kind instantiate the kind's essence and *only* members of the kind do so. Commonly, such kind essences are conceived of in terms of intrinsic properties of things (e.g., Aristotelian substantial forms, inner principles of existence and activity, material structure, material composition). Traditional examples of such natural kinds include the kinds of elementary particles, the chemical elements and isotopes and biological species of organisms (e.g., Bird and Tobin 2010). In present-day philosophy, this view of natural kinds continues to be endorsed, in particular among analytic metaphysicians, philosophers of language (especially those continuing Kripke's and Putnam's work on reference theory), and philosophers of science who defend strong versions of scientific realism.[9] Among many philosophers of science, however, the tradition has waned.

While criticisms of the tradition already appeared in the 1950s (an important such criticism was Nelson Goodman's discussion of relevant kinds and projectible predicates – e.g., Goodman 1954), an important step in this respect was taken in the 1960s–1970s in the philosophy of biology, when it was noticed that a traditional example of natural kinds – biological species – did not fit into the essentialist framework. What could the kind essences of species be? It seemed obvious that the essential intrinsic properties of organisms should be their genes (as the underlying causes of organisms' phenotypic features) and that accordingly species as kinds were to be

[8] Elsewhere, I have given other reasons to be skeptical about metaphysics-oriented approaches to developing a theory of natural kinds (Reydon 2010).

[9] Among the most forceful contemporary defenders of scientific realism and natural kind essentialism are Ellis (2001, 2002), Oderberg (2007), and Devitt (2008, 2010).

defined by genetic essences.[10] However, not only had biology so far failed to identify any species-specific genetic essences (and, moreover, had never aimed to do so), but there also were good theoretical reasons to think that species don't have genetic essences. Species, after all, are subject to evolutionary change and open-ended genetic variation between the organisms of a species is not only the outcome of evolution but also a necessary requirement for evolution to occur.

These considerations led biologist Michael Ghiselin and philosopher David Hull to suggest an alternative view of the nature of species that could avoid the problems encountered by the view that species were essentialist natural kinds. Ghiselin and Hull argued that the failure of biologists to uncover the one true nature of *Homo sapiens* or the intrinsic essence that characterizes all and only *Arabidopsis thaliana* plants is due to a simple fact: there is nothing there to be found! According to Ghiselin and Hull, species are not to be understood ontologically as classes or kinds but as individuals (Ghiselin 1966, 1974; Hull 1976, 1978). That is, a species is not a kind with particular organisms as its members, but itself a concrete particular entity with organisms as its constituent parts. Species at most have individual essences but not kind essences (Okasha 2002). Ontologically, then, biological species should be grouped together with organisms, firms, and material objects, not with the chemical elements and the various kinds of elementary particles in the Standard Model. If this is correct, it was only to be expected that the quest for organismal properties (genetic or otherwise) that would constitute the kind essences of the species that organisms belonged to would be in vain: species just aren't the sort of things that have essences of the sort that natural kinds are supposed to have.[11]

Ghiselin's and Hull's suggestion apparently resolved the question about species essences and today most biologists and philosophers of biology seem to agree that species metaphysically belong to the category of individuals not natural kinds. However, one problem persisted: species names, such as "*Arabidopsis thaliana*," often function as kind terms in biological reasoning. Biologists commonly refer to species when making generalizations about groups of organisms and inferring from observations on a few members of the species to claims about all members of the species. As Millikan, for example, pointed out: "Inductions from one member of a species to the next often hold up for very good reason. Were this not so, there could be no science of biology" (Millikan 2000, p. 208). This epistemic function is a characteristic role of natural kinds.[12] The case of species thus presented a problem: species cannot be conceived of as natural kinds under traditional essentialist accounts,

[10] A suggestion that appears in Kripke's and Putnam's work.

[11] However, there is a movement to return to essentialism about species – see Oderberg (2007), Devitt (2008, 2010), or the recent symposium on "The New Biological Essentialism" at the 21st biennial meeting of the Philosophy of Science Association in 2008 (for the four published papers of the symposium, see *Philosophy of Science*, volume 77, issue no. 5, 2010, pp. 648–701).

[12] On many accounts, this characteristic is explicated by connecting natural kinds to laws of nature. The claim then is that there are laws about all natural kinds and those kinds that feature in laws of nature are precisely the natural kinds that exist (Reydon 2010).

but nevertheless they perform a characteristic epistemic role of natural kinds. The question is how these two facts of the matter might be reconciled.

Recently, a number of philosophers have begun to address this question (most importantly, Boyd (1999a, b, 2000, 2010), Griffiths (1999), Millikan (1999a, b, 2000), Wilson (1999a, b, 2005), Keller et al. (2003), Brigandt (2009), Wilson et al. (2007)). According to these authors, because in biological reasoning species perform a characteristic epistemic role of natural kinds, they should be conceived of as natural kinds after all. In this approach to the problem of natural kinds, epistemology is more important than metaphysics: what distinguishes natural kinds from other sorts of groupings of things are their epistemic roles, rather than meeting particular metaphysical criteria such as being associated with kind essences.

Presumably, an important motivation behind this approach was the persistent failure of traditional accounts, which focused on assumed metaphysical criteria for natural kindhood, to yield a theory of natural kinds able to account for all the kinds that featured in the various sciences. Most theories accounted only for a few special cases and ignored most of the kinds that scientists actually used, using a priori metaphysical assumptions to separate "good" natural kinds from other sorts of groupings and resulting in the recognition of only a small group of kinds as natural kinds (Reydon 2010; see also Churchland 1985; Hacking 1991, 2007). A natural move thus was to reconceive of the problem of natural kinds as not being about what kinds of things the "furniture of the world" is made up from but about the nature of the kinds of things that humans refer to in reasoning and that investigations can provide knowledge about. As Boyd put it: "[i]t is a truism that the philosophical theory of natural kinds is about how classificatory schemes come to contribute to the epistemic reliability of inductive and explanatory practices" (Boyd 1999a, p. 146; 2000, pp. 55–56). And: "the theory of natural kinds *just is (nothing but)* the theory of how accommodation is (sometimes) achieved between our linguistic, classificatory and inferential practices and the causal structure of the world" (Boyd 2000, p. 66; emphasis added).

Basically, the approach taken by the abovementioned authors inverts the order of importance of different kinds of criteria for natural kindhood and in doing so redefines the problem from (previously) a metaphysical one into (now) an epistemological one. The principal criteria for being a natural kind used to be metaphysical: a kind is a natural kind if and only if it really exists in the world (whatever it may mean for kinds to exist), independently of human consciousness, human interests, and human practices, and is associated with a particular kind essence (however one exactly conceives of kind essences). On the alternative approach the principal criteria for being a natural kind no longer are metaphysical but epistemological: what counts is being useful in human epistemic practices, such as inference and explanation, by corresponding in some way (which is still to be explicated) to the state of affairs in nature. On this new approach, the metaphysical project of clarifying in what way the kinds referred to in epistemical practices reflect the state of affairs in nature comes into focus only *after* the kinds have been individuated on epistemological grounds. In this respect, this approach can be seen as exemplary of what I call an "epistemological turn" in thinking about kinds: the principal criteria for distinguishing natural kinds from other sorts of groupings are epistemological, while metaphysical issues come second in line, leading to the recognition of many more natural kinds than was the case on metaphysics-oriented approaches.[13]

As on this view natural kinds are epistemically successful kinds – a given grouping of things is a natural kind if and only if it features successfully in human epistemic practices because of its being anchored in some way in nature – the principal question for a theory of natural kinds is what epistemic success consists in and in what ways reference to kinds contributes in realizing it. But reference to kinds can only be epistemically successful if there is something in the world that underwrites these epistemic practices. Thus, there is a metaphysical aspect to the problem too, namely, to explicate what about the world "out there" epistemically successful kinds refer to. The point is that this metaphysical issue comes second in line, after the epistemology of kinds has been sufficiently clarified.

Boyd attempted to resolve these issues by means of what has come to be known as the Homeostatic Property Cluster theory of kinds (henceforth, HPC theory), about which I shall have more to say in the following section.[14] At this point, I should only point out that by making natural kinds dependent on human epistemic practices HPC theory avoids the problem that was highlighted in Sect. 8.2. From the HPC perspective, there is no need to conceive of natural kinds as being necessarily mind-independent and to distinguish them from mind-dependent artificial kinds. On the contrary, natural kinds are mind-dependent groupings too, as they crucially depend on human epistemic practices.[15] In this respect,

---

[13] While in the philosophies of the special sciences the turn can be located in the 1980s–1990s, with the elaboration and growing acceptance of Boyd's theory of kinds, in analytic philosophy more generally the turn probably lies with Nelson Goodman's insistence on kinds as the extensions of projectible predicates (Goodman 1954, 1984, p. 21; Boyd 1999a, p. 147; Griffiths 1999, p. 215). One might locate its roots at a much earlier time, for example, in mid-nineteenth century British Empiricism – where Hacking (1991, 2007) locates the origin of the philosophy of natural kinds – or even further back in time in Locke's *Essay*. I have briefly addressed this issue elsewhere (Reydon 2010) but have to leave the historiography of the epistemological turn for future work.

[14] For details, see Boyd (1999a, b, 2000, 2010), Griffiths (1999), Wilson (1999a, b, 2005), Keller et al. (2003), and Wilson et al. (2007). My discussion of HPC theory in Sect. 8.4 is based on the discussion I provided in Reydon (2009).

[15] As one of my coeditors of this volume pointed out, two notions of mind dependency should be distinguished here: human intentions play a role in defining kinds *within* particular epistemic contexts and human intentions underlie these epistemic contexts themselves. In the case of natural kinds as mind-dependent groupings, both kinds of mind dependency are involved. In the traditional contrast between mind-independent natural kinds and mind-dependent artificial kinds, too, both notions of mind dependence were involved but not clearly distinguished. There, the central idea was that for mind-independent kinds nature provided a uniquely privileged basis for grouping things into kinds such that – even though the grouping of things into kinds is done by humans in the context of various epistemic practices – the outcome of grouping practices (when done correctly) does not depend on human intentions. For artificial kinds, no such uniquely privileged basis is available (either because it doesn't exists or because we in principle don't have epistemic access to it), such that the grouping of things into kinds always crucially depends on human interests.

the epistemology-oriented approach toward the problem of natural kinds might after all open up prospects to develop a theory of artifact kinds by taking recourse to an available theory of natural kinds.

8.4 Artifact Kinds as HPC Kinds?

To what extent are the hopes that HPC theory could account for artifact kinds or even that a unified account of natural and artifact kinds might be within reach by applying HPC theory to both kinds of kinds warranted?

HPC theory begins from the observation that most kinds used in the various special sciences do not collect things that are in every which way the same. Rather, many sciences use kinds of which the members are very similar to one another in theoretically important ways even though they vary in numerous respects. Furthermore, these kinds typically rest on the assumption that the members of a kind exhibit largely similar properties due to largely similar causes: they are alike because similar causes (have) operate(d) on them (Boyd 1999a, pp. 142–144). Accordingly, HPC theory assumes that natural kindhood and kind membership cannot be understood in terms of separately necessary and jointly sufficient essential properties that are exhibited without exception by all and only the members of the kind. Rather, the cluster of properties that are found to regularly, but not exceptionlessly, occur together in the members of putative kinds should come into play here. But if for a given natural kind there is no set of properties unique to and characteristic of all members of that kind, kinds cannot be individuated by property clusters alone: if there is considerable variation between the putative members of a kind, the relevant property cluster can only be identified after kind membership has been established comparatively well. Accordingly, HPC theory adds a second element to the definition – a set of causal factors that underlie the observed property clustering – and takes the combination of these two elements to uniquely determine a kind: a kind is determined by the properties that are found to repeatedly occur together in its members plus the underlying factors that cause this clustering. (Boyd calls these causal factors "homeostatic mechanisms" and emphasized that the term should not be read too literally.)

In order to do justice to the state of affairs in the world in which entities are hardly ever exactly alike, HPC theory conceives of the two elements of the definition of a kind in an open-ended manner. No property or combination of properties is necessarily unique to one property cluster, the property cluster associated with a kind may in time come to include new properties, and present properties may cease to be exhibited by members of the kind. Similarly, no causal factor or combination of factors is necessarily unique to the set of "homeostatic mechanisms" associated with a kind, and causal factors may begin or cease to operate on the members of the kind. If there were one fixed characteristic set of properties for every kind, HPC theory would merely be a form of traditional kind essentialism (albeit with less strict essences than on the traditional view) and as such not be able to avoid the problems that confront essentialism. But by conceiving of property clusters and sets of "homeostatic mechanisms" as open-ended, HPC theory aims to constitute an account of natural kinds that is sufficiently flexible to accommodate all the various kinds that feature in the various special sciences, as well as the traditionally recognized natural kinds.

Consider, for example, the case of biological species. Species are products of evolution and during their existence subject to ongoing, open-ended evolution. Newly evolved traits can come to be widespread and old traits can be lost as time goes by, while there is no reason to assume that any particular core set of traits will be conserved throughout the entire species' lifetime – other than developmentally deeply entrenched traits that are conserved over evolutionary timescales much longer than the species' lifetime, that is.[16] Furthermore, in the case of a speciation event in which a new species branches off from its ancestor species, the member organisms of the two species will often continue to be characterized by the same family of properties for quite some time. The same holds for the causes underlying the presence of organismal traits. Traits that are deeply genetically entrenched will remain present for very long times, whereas not very deeply entrenched traits can cease to be present with a species' members before the species itself has ceased to exist. If the relevant causal factors are environmental, the environment may change heavily during a species' lifetime without speciation occurring or remain the same over the lifetimes of an ancestral species and a series of its descendants. Therefore, in order to be able to conceive of species as natural kinds of organisms, as Boyd (1999a, b) does, the theory of kinds must be sufficiently flexible and allow open-ended change in the property sets and sets of underlying mechanisms that are taken as determining kinds.

Elsewhere (Reydon 2009), I have suggested that this flexibility constitutes both a strength and a fatal weakness of HPC theory. In my view, HPC theory fails as an account of natural kinds because it does not actually provide any criteria for kind membership or for distinguishing between natural kinds and other sorts of groupings. By individuating kinds in terms of property clusters plus the causal factors that underlie this clustering, HPC theory only explicates why members of a *given* kind exhibit similar properties, such that reference to kinds can successfully ground inferences, generalizations, etc. But HPC theory does not explicate what makes a given thing a member of one kind rather than another and, accordingly, fails to fix the boundaries of kinds. Simply exhibiting many of the properties that recognized members of the kind possess won't do, nor will exhibiting these properties due to causes similar to those that caused them to be present in recognized kind members. For the property clusters and sets of "homeostatic mechanisms" that in HPC theory determine, kinds are open-ended to such an extent that – given enough time – at

[16] Many organismal traits are conserved over evolutionary timescales far extending the lifetimes of individual species. Consider, for example, the presence of a backbone in the Vertebrata. This trait occurs in all members of a species of vertebrates – all viable members of *Canis lupus* have a backbone – but it is not typical for one particular species and thus cannot serve to distinguish the species as a kind from the many other species in which the trait occurs too (Reydon 2006).

different times the same kind may be characterized by wholly different property clusters and underlying causes (making exhibiting the same properties for the same reasons not necessary for kind membership), while in principle (even though probably not often in practice) different kinds may in time come to be characterized by similar property clusters and underlying causes (making exhibiting the same properties for the same reasons not sufficient for kind membership).

According to HPC theory, natural kinds are not simply found "out there" in the world, nor do they emerge (as artificial kinds do) from any which way of classifying things we might find suitable to our purposes. Rather, as Boyd (1999a, 2000) – to my mind rightly – emphasized, natural kinds emerge from human interactions with nature in epistemic practices – in practices of gathering knowledge, explaining observed phenomena, predicting future events, etc. It follows that any theory of kinds should refer not to those properties of things that make them members of particular kinds, *period*, but to those properties that make things members of kinds *within some particular epistemic context*. Thus, the context under consideration determines which of the myriad properties of things determine kind membership and which are irrelevant in this respect.

However, while this is one of the principal insights underlying HPC theory, the relation of kinds to epistemic contexts is not part of the theory itself but is simply taken as a given in that HPC theory presupposes the definitions of the kinds to which the theory is applied to be provided by the relevant epistemic context. This means that HPC theory only accounts for one aspect of natural kinds, namely, their epistemology, but not for the metaphysics of individual kinds as it is left to the various scientific disciplines (and other epistemic contexts) to explicate the basis for the classification of their subject matter into kinds.[17] Scientists do not simply group things into kinds because they happen to have similar properties, but they group things on theoretical grounds. That is, scientists select those properties that they deem relevant against the background of the theoretical framework they use and group the things under study accordingly. (Often, these will be properties that feature in the explanations provided by a field of investigation.) This is illustrated by the classification of organisms: organisms can be similar in very many respects, but the similarities that count when grouping organisms into species and higher taxa are those that indicate closeness of ancestry (Darwin 1859, ch. XIII). Accordingly, Boyd and other proponents of HPC theory are concerned with kinds that feature in the various sciences, reaching from the physical sciences via the biological sciences to cognitive science, social science, and even the humanities. These epistemic contexts provide the groupings of things into kinds that in turn can be explicated in more detail by applying HPC theory to them.

[17] HPC theory specifies at most a very superficial metaphysics of kinds. Elsewhere, I have suggested that precisely herein a possibility might lie for turning HPC theory into a full-blown theory of kinds that accounts for the epistemology *and* metaphysics of kinds: adding the factor that actually makes a kind epistemically important in a particular context as a third element to the two-part HPC definition of a kind allows the HPC definition to explain epistemic success as well as the metaphysical aspects of kind membership (Reydon 2009).

A similar situation obtains for artifacts: artifacts can be similar in numerous respects, but not all similarities have equal weight when grouping artifacts into kinds that serve the purposes of particular epistemic contexts. For the case of artifact kinds this brings up the question what might be the relevant epistemic contexts to examine when elaborating a theory of artifact kinds. One possibility is to look at academia – in the same way as proponents of HPC theory usually consider kinds in the various sciences – and take the various academic disciplines in which kinds of artifacts feature as the relevant contexts for a theory of artifact kinds. On this option, one should examine the kinds and classifications featuring in the sciences of artifacts, such as the various engineering disciplines found at polytechnic institutes, technical universities, and technical research institutes, as well as perhaps other fields that study artifacts (e.g., anthropology, archaeology, and museum studies). However, many if not most artifacts are conceived, designed, and examined not within academia but in such places as architects' offices, research and development departments of companies, etc., such that one might have to broaden the domain under consideration. But there is no a priori reason why one should limit the relevant epistemic contexts for theories of artifact kinds to this broader domain or even to take academic and other professional engineering and design contexts as relevant. One might also take "folk technology" and "folk" classifications of artifacts as the principal relevant context (Preston 2013).[18]

Depending on what one takes as the relevant epistemic contexts, different epistemic goals will move into focus. Probably the engineering disciplines and the non-academic design and manufacturing practices (and even "folk technology") share many epistemic goals with the natural, life, and social sciences (e.g., achieving a body of reliable knowledge, realizing stable predictions, having a means to interfere with and control processes), but it is unlikely that they have *exactly* the same epistemic aims as are traditionally found in the sciences. At the very least, explanation and understanding often seem to come in different modes in engineering and design contexts on the one hand and in the natural sciences on the other hand. While the sciences strive to explain and/or understand given phenomena that hitherto have gone unexplained, in engineering and design, some level of understanding of the artifact must *precede* its existence as one needs some understanding of how the relevant phenomena work before one can design and make an artifact that is partly based on these phenomena. In addition, explanations here often have different targets from the scientific target (to explain regularities that are found in the world but aren't yet understood), such as explaining malfunction or the occurrence of unintended consequences (e.g., Pitt 2000, pp. 41–51).

I cannot elaborate on this issue here. A detailed analysis of the epistemic goals of engineering and design in comparison to the sciences is, to my knowledge, still unavailable, but it is at least prima facie plausible that, in engineering and design

[18] Of course the same holds for natural kinds. Even though analytic metaphysicians and philosophers of science typically consider the theory of natural kinds in relation to those kinds that feature in the sciences, one might also take "folk" kinds and classifications as the relevant context for studying natural kinds. Indeed, cognitive psychologists often do (e.g., Keil 1989).

contexts, partly different epistemic aims play important roles from the ones that are central in the physical, life, and social sciences. This is not to say that the epistemic contexts in which artifact kinds feature are completely different from the ones in which natural kinds come into play; there will be overlap. My point is merely that when looking at artifact kinds, some of the important epistemic roles that they play are likely to be different from the ones that natural kinds play in the contexts in which these feature.[19]

The upshot is that for HPC theory or other epistemology-oriented approaches to be applied to the case of artifact kinds, the relevant epistemic contexts and the epistemic functions of referring to artifact kinds within these contexts need to be determined first. After all, on this approach, kinds are individuated by the reference that we make to them in successful generalizations, explanations, predictions, etc. In the case of natural kinds, philosophers often take the various sciences as the relevant epistemic contexts and think of natural kinds as just those kinds referred to in scientific reasoning. But, as pointed out above, for both natural and artifact kinds we seem to have a choice with respect to which epistemic contexts are deemed the relevant ones. Much of this choice, it seems to me, depends on what we are willing to consider as belonging into the general categories of "artifact" and "artifact kind" in the first place. We will only be able to examine the epistemic roles of reference to artifact kinds if we know at which kinds to look. Thus, it needs to be decided first about kinds *of what* we are seeking a theory when attempting to develop an account of artifact kinds. But here any project to elaborate an account of artifact kinds runs into problems, as there seems to be no general agreement among the specialists in the field about where exactly the boundaries of the category of artifacts lie.

As a first approximation, philosophers commonly think of artifacts as man-made objects (e.g., Franssen 2008, pp. 21–22; Franssen et al. 2009, section 2.5; Verbeek and Vermaas 2009, p. 165). Note, however, that having been made by people or being the product of human action cannot be sufficient to delimit the artifact category. Members of paradigmatic natural kinds can be made intentionally by humans too: the Higgs bosons that scientists hope to find in the Large Hadron Collider experiments at CERN constitute one example; the ultraheavy chemical elements created in laboratory setups are another (although here "making" presumably doesn't have precisely the same sense as when one talks about the making of more paradigmatic artifacts such as hammers or laptop computers). Conversely, artifacts can also be naturally occurring objects that are selected by humans for a particular use. Thus, an appropriately shaped and sized stone used as a paperweight or a suitably large seashell used as an ashtray are sometimes seen as artifacts on a par with hammers or houses. Accordingly, philosophers of technology often employ a notion of "artifact" that covers objects *made* for a particular use as well as "objects that are intentionally or less intentionally *selected* to be used" (Verbeek and Vermaas 2009, pp. 165–166, emphasis added; see also Sperber 2007, pp. 124–125, for discussion),

[19] But see Bunge (1966) for a view that the sciences and engineering disciplines actually *are* epistemically the same sort of endeavor.

although others prefer to place objects of the latter sort into the separate category of "naturefacts" (Oswalt 1973) in order to distinguish them from man-made things.

In philosophy of technology, however, not all made or selected objects are necessarily counted as artifacts: often the notion of "artifact" is limited to apply only to objects that are made/selected in order to serve a particular purpose. Things that were made or selected without having been meant by their makers/selectors to serve a specific purpose – such as unintended byproducts and waste products of production processes – tend not to be counted as artifacts (Franssen 2008, p. 22; Franssen et al. 2009, section 2.5; Verbeek and Vermaas 2009, p. 165). An additional restriction often imposed by philosophers of technology is that in order to count as an artifact, an entity should successfully *realize* or at least be *capable of realizing* the purpose that its maker had in mind when making it: if the maker fails in realizing his/her intentions, he/she hasn't produced an artifact but only "scrap" (Hilpinen 2008, section 3; Verbeek and Vermaas 2009, pp. 165–166). The ideas in the minds of making subjects thus play a crucial role in determining whether a given object is an artifact and if so, of what kind.[20]

But this strict view of what it is to be an artifact a priori excludes some categories of man-made things from the category of artifacts that arguably should be included when trying to elaborate a philosophical theory of artifact kinds. For example, practices of classifying waste products and byproducts of industrial production processes are as ubiquitous and as important in everyday life as well as in the engineering disciplines as are practices of classifying "proper" artifacts. What could be a good reason to exclude kinds of non-purposefully created things like byproducts and waste from the realm of artifact kinds that is to be covered by the desired philosophical theory, focusing only on kinds of purposefully made objects?

The same seems to hold for things of which the purpose isn't clear or is very difficult to specify, such as works of art or social institutions. Often the purpose of a work of art is unclear, but still there is much to say for counting works of arts as artifacts (e.g., Levinson 2007). And indeed, there are academic disciplines that study such artifacts and classify them: archaeology, art history, and cultural anthropology are among them. Furthermore, consider "things" that aren't immediately tangible, material objects – such as the Internet, the Coca-Cola Company, or the European Union. For these sorts of things, too, there are sciences that study and classify them. Are these artifacts? Often, social institutions, firms, and "things" like the Internet are designed and created with well-specified purposes in mind but later assume a life of their own and serve different purposes than the ones they were originally created for. Still, they clearly aren't naturally occurring objects but things intentionally created by humans and thus exhibit an important characteristic of artifacts. It seems strange not to count them as artifacts. Similarly, it is unclear why "scrap" and objects made or selected by animals should a priori be excluded from the artifact category. Indeed, some authors use a notion of artifact, according to

[20] Cf. Thomasson's claim that making an artifact of kind *K* must involve having a "substantive concept of the nature of *K*s that largely matches that of some group of prior makers of *K*s" (2003, 600; cf. 2007, 60–63; Chap. 4, this volume).

which the maker must accept the entity as something that *might have realized* his/her intentions, thus including defective artifacts. And if natural objects selected by humans to serve particular purposes are counted as artifacts, why not count birds' nests, termite hills, beaver dams, or the twigs and leaves that animals use as tools (e.g., Gould 2007)?

In all likelihood, it will be easier to achieve a theory that explicates the epistemology and metaphysics of technical artifact kinds – artifact kinds that feature in the engineering disciplines – than to develop a theory of artifact kinds that also accounts for byproducts, scrap, art works, social institutions, etc. Within the context of the philosophy of the engineering sciences one would probably only consider kinds of technical artifacts, in the same way as in the philosophy of science discussions on natural kinds are often limited to scientific kinds. But there is no a priori reason to limit one's considerations to kinds that feature in the established academic engineering disciplines while not also looking at kinds featuring in design and manufacturing practices outside academia (and even then one will presumably not have covered the entire domain of the artifactual). At any rate, a question that needs to be answered prior to attempting to devise a theory of artifact kinds inspired by HPC theory – or along the lines of any other epistemology-oriented approach to kinds – is how broad the scope of application of the desired theory should be and which epistemic and/or pragmatic contexts are counted as relevant.

8.5 Outlook

In the preceding sections, I have argued that traditional metaphysics-oriented approaches to elaborating a theory of kinds are on the wrong track, both when it comes to natural kinds and artifacts kinds, but that there has been an epistemological turn in the philosophy of kinds that looks more promising. The approach I favored involves addressing the epistemology of kinds before trying to elucidate their metaphysics. I have examined to some extent whether the most popular epistemology-oriented account of natural kinds, HPC theory, might be applied to artifact kinds and argued that some issues that arise in this context deserve attention:

1. A general problem faced by HPC theory, namely, that HPC theory fails to provide membership criteria for kinds
2. The question which epistemic roles references to artifact kinds perform
3. The question which epistemic contexts are relevant when examining the epistemic roles of reference to artifact kinds
4. The question how wide the scope of application of the desired account of artifact kinds should be, i.e., which things should be counted as artifacts and which should not

While the occurrence of these issues might be taken to suggest that epistemology-oriented approaches to developing a theory of artifact kinds will fare no better than metaphysics-oriented approaches, I think the former still are in a better position than the latter. For one, metaphysics-oriented approaches tend to be more limited than epistemology-oriented approaches in their attempts to yield a unified account of natural and artifact kinds, as they often (but, to be sure, not necessarily always) invoke a priori criteria for what it is to be a natural kind – for instance, the traditional principle that mind-dependent kinds are fundamentally different from natural kinds that supposedly exist wholly independently from human cognition, or the principle that natural kinds are to be conceived of in terms of kind essences. Such a priori criteria typically are difficult to reconcile with the kinds humans actually refer to (cf. Hacking 1991, 2007; Reydon 2010), a prominent example (discussed above) being the case of biological species. By looking at epistemic rather than metaphysical criteria for what it is to be a natural kind and for how things should be allocated to kinds, epistemology-oriented approaches are in a better position to avoid this difficulty.[21]

But to my mind the most promising aspect of epistemology-oriented approaches is that they leave open the possibility to fill in the metaphysics of different kinds in different ways because kinds are individuated on epistemological grounds. Their metaphysics then is to be filled in on a case-by-case basis, leaving open what the metaphysics of the various sorts of kinds will look like and whether in the end a unified metaphysics for all kinds is achievable. In the case of artifact kinds, this feature of HPC theory and similar approaches is especially important, because of the variety of extant views of what the nature of artifacts consists in and what makes a given artifact the kind of thing that it is. The straightforward way is to characterize artifacts by their functions (e.g., Kornblith 1980, p. 112). However, it is now widely acknowledged that an artifact's function cannot fully determine its kind membership (Franssen et al. 2009, section 2.5). Better options might be to characterize them by functions plus other features (Carrara and Vermaas 2009) such as operational principles, by human actions involving artifacts in the contexts of use plans (Houkes and Vermaas 2004, 2010), or by means of a dual nature combining structure and function (Kroes and Meijers 2006; Kroes 2010). In this respect, epistemology-oriented approaches have an advantage over metaphysics-oriented ones, in that they don't need to wait for the discussion on the nature of artifacts to be decided before addressing the question of artifact kinds.

Any metaphysics-oriented approach to developing a theory of artifact kinds must begin by agreeing upon the *kind of metaphysics* that is sought – a metaphysics in terms of necessary and sufficient properties for kind membership, one that recognizes only structures as real, one that allows both non-sharply delimited kinds next to strict kinds, one that conceives of artifacts as being individuated only by functions, one that thinks of artifacts as having a dual nature, etc. Thus, on metaphysics-oriented approaches, as long as the nature of artifacts is an unsettled issue, the nature of artifact kinds must remain open too. Epistemology-oriented approaches, in contrast, can begin to explicate the epistemology of artifact kinds in the various contexts in which they feature, explore whether a unified epistemology

---

[21] But see Elder (2007; Chap. 3, this volume) for an attempt at devising an account of mind-independent essences for artifact kinds that perhaps could avoid these difficulties.

is possible that covers all cases, and then go on to see whether an overarching metaphysics of artifact kinds is feasible or a pluralist metaphysics is required.

These advantageous features of epistemology-oriented approaches in the case of artifact kinds in turn provide additional support for the epistemological turn that is on its way in the philosophy of natural kinds.

Acknowledgments I thank Maarten Franssen, Peter Kroes, and Pieter Vermaas for their kind invitation to participate in the workshop "Engineering Artefact Kinds" (Delft, June 2009) and for accepting my contribution for the ensuing volume before I joined them later as the book's fourth editor. I am also indebted to them as well as to Beth Preston and an anonymous reviewer for providing helpful comments and criticisms on earlier versions of this chapter.

References

Bird, A., & Tobin, E. (2010). Natural kinds. In E. N. Zalta (Ed.), *Stanford encyclopedia of philosophy* (Summer 2010 ed.). http://plato.stanford.edu/archives/sum2010/entries/natural-kinds/

Boyd, R. N. (1999a). Homeostasis, species, and higher taxa. In R. A. Wilson (Ed.), *Species: New interdisciplinary essays* (pp. 141–185). Cambridge, MA: MIT Press.

Boyd, R. N. (1999b). Kinds, complexity and multiple realization. *Philosophical Studies, 95*, 67–98.

Boyd, R. N. (2000). Kinds as the "workmanship of men": Realism, constructivism, and natural kinds. In J. Nida-Rümelin (Ed.), *Rationalität, Realismus, Revision: Vorträge des 3. internationalen Kongresses der Gesellschaft für Analytische Philosophie* (pp. 52–89). Berlin: Walter de Gruyter.

Boyd, R. N. (2010). Homeostasis, higher taxa, and monophyly. *Philosophy of Science, 77*, 686–701.

Brigandt, I. (2009). Natural kinds in evolution and systematics: Metaphysical and epistemological considerations. *Acta Biotheoretica, 57*, 77–97.

Bunge, M. (1966). Technology as applied science. *Technology and Culture, 7*, 329–347.

Bunge, M. (1977). *Treatise on basic philosophy. Vol. 3: Ontology I: The furniture of the world.* Dordrecht: Reidel.

Carrara, M., & Vermaas, P. E. (2009). The fine-grained metaphysics of artifactual and biological functional kinds. *Synthese, 169*, 125–143.

Churchland, P. M. (1985). Conceptual progress and word/world relations: In search of the essence of natural kinds. *Canadian Journal of Philosophy, 15*, 1–17.

Darwin, C. R. (1859). *On the origin of species by means of natural selection, or the preservation of favoured races in the struggle for life*. London: John Murray.

Devitt, M. (2008). Resurrecting biological essentialism. *Philosophy of Science, 75*, 344–382.

Devitt, M. (2010). Species have (partly) intrinsic essences. *Philosophy of Science, 77*, 648–661.

Elder, C. L. (2007). On the place of artifacts in ontology. In E. Margolis & S. Laurence (Eds.), *Creations of the mind: Theories of artifacts and their representation* (pp. 33–51). Oxford: Oxford University Press.

Ellis, B. (2001). *Scientific essentialism*. Cambridge: Cambridge University Press.

Ellis, B. (2002). *The philosophy of nature: A guide to the new essentialism*. Montreal/Kingston: McGill-Queen's University Press.

Ereshefsky, M. (2001). *The poverty of the Linnaean hierarchy: A philosophical study of biological taxonomy*. Cambridge: Cambridge University Press.

Franssen, M. P. M. (2008). Design, use, and the physical and intentional aspects of technical artifacts. In P. E. Vermaas, P. Kroes, A. Light, & S. A. Moore (Eds.), *Philosophy and design: From engineering to architecture* (pp. 21–35). Dordrecht: Springer.

Franssen, M. P. M., Lokhorst, G.-J., & van de Poel, I. (2009). Philosophy of technology. In E. N. Zalta (Ed.), *Stanford encyclopedia of philosophy* (Fall 2009 ed.). http://plato.stanford.edu/archives/fall2009/entries/technology/

Ghiselin, M. T. (1966). On psychologism in the logic of taxonomic controversies. *Systematic Zoology, 15*, 207–215.

Ghiselin, M. T. (1974). A radical solution to the species problem. *Systematic Zoology, 23*, 536–544.

Goodman, N. (1954). *Fact, fiction, and forecast*. London: Athlone.

Goodman, N. (1984). *Of mind and other matters*. Cambridge, MA: Harvard University Press.

Gould, J. L. (2007). Animal artifacts. In E. Margolis & S. Laurence (Eds.), *Creations of the mind: Theories of artifacts and their representation* (pp. 249–266). Oxford: Oxford University Press.

Griffiths, P. E. (1999). Squaring the circle: Natural kinds with historical essences. In R. A. Wilson (Ed.), *Species: New interdisciplinary essays* (pp. 209–228). Cambridge, MA: MIT Press.

Hacking, I. (1991). A tradition of natural kinds. *Philosophical Studies, 61*, 109–126.

Hacking, I. (2007). Natural kinds: Rosy dawn, scholastic twilight. In A. O'Hear (Ed.), *Philosophy of science* (Philosophy Vol. 82, Royal Institute of Philosophy Supplement 61, pp. 209–228). Cambridge: Cambridge University Press.

Hilpinen, R. (2008). Artifact. In E. N. Zalta (Ed.), *Stanford encyclopedia of philosophy* (Fall 2008 ed.). http://plato.stanford.edu/archives/fall2008/entries/artifact/

Houkes, W., & Vermaas, P. E. (2004). Actions versus functions: A plea for an alternative metaphysics of artifacts. *The Monist, 87*, 52–71.

Houkes, W., & Vermaas, P. E. (2010). *Technical functions: On the use and design of artefacts*. Dordrecht: Springer.

Hull, D. L. (1976). Are species really individuals? *Systematic Zoology, 25*, 174–191.

Hull, D. L. (1978). A matter of individuality. *Philosophy of Science, 45*, 335–360.

Keil, F. C. (1989). *Concepts, kinds, and cognitive development*. Cambridge, MA: MIT Press.

Keller, R. A., Boyd, R. N., & Wheeler, Q. D. (2003). The illogical basis of phylogenetic nomenclature. *The Botanical Review, 69*, 93–110.

Kornblith, H. (1980). Referring to artifacts. *Philosophical Review, 89*, 109–114.

Kroes, P. (2010). Engineering and the dual nature of technical artefacts. *Cambridge Journal of Economics, 34*, 51–62.

Kroes, P., & Meijers, A. W. M. (2006). The dual nature of technical artifacts. *Studies in History and Philosophy of Science, 37*, 1–4.

Levinson, J. (2007). Artworks as artifacts. In E. Margolis & S. Laurence (Eds.), *Creations of the mind: Theories of artifacts and their representation* (pp. 74–82). Oxford: Oxford University Press.

Losonsky, M. (1990). The nature of artifacts. *Philosophy, 65*, 81–88.

Margolis, E., & Laurence, S. (Eds.). (2007). *Creations of the mind: Theories of artifacts and their representation*. Oxford: Oxford University Press.

McOuat, G. (2009). The origins of natural kinds: Keeping "essentialism" at bay in the age of reform. *Intellectual History Review, 19*, 211–230.

Millikan, R. G. (1999a). Historical kinds and the "special sciences". *Philosophical Studies, 95*, 45–65.

Millikan, R. G. (1999b). Response to Boyd's commentary. *Philosophical Studies, 95*, 99–102.

Millikan, R. G. (2000). *On clear and confused ideas: An essay about substance concepts*. Cambridge: Cambridge University Press.

Oderberg, D. S. (2007). *Real essentialism*. New York/London: Routledge.

Okasha, S. (2002). Darwinian metaphysics: Species and the question of essentialism. *Synthese, 131*, 191–213.

Oswalt, W. H. (1973). *Habitat and technology: The evolution of hunting*. New York: Holt, Rinehart, and Winston.

Pitt, J. C. (2000). *Thinking about technology: Foundations of the philosophy of technology*. New York/London: Seven Bridges Press.

Preston, B. (2013). Ethnotechnology: A manifesto. In M. Franssen, P. Kroes, T. A. C. Reydon, & P. E. Vermaas (Eds.), *Artefact kinds: Ontology and the human-made world* (pp. 145–163). Heidelberg: Springer.

Reydon, T. A. C. (2006). Generalizations and kinds in natural science: The case of species. *Studies in History and Philosophy of Biological and Biomedical Sciences, 37*, 230–255.
Reydon, T. A. C. (2009). How to fix kind membership: A problem for HPC-theory and a solution. *Philosophy of Science, 76*, 724–736.
Reydon, T. A. C. (2010). Natural kind theory as a tool for philosophers of science. In M. Suárez, M. Dorato, & M. Rédei (Eds.), *EPSA – Epistemology and methodology of science: Launch of the European Philosophy of Science Association* (pp. 245–254). Dordrecht: Springer.
Sperber, D. (2007). Seedless grapes: Nature and culture. In E. Margolis & S. Laurence (Eds.), *Creations of the mind: Theories of artifacts and their representation* (pp. 124–137). Oxford: Oxford University Press.
Thomasson, A. L. (2003). Realism and human kinds. *Philosophy and Phenomenological Research, 67*, 580–609.
Thomasson, A. L. (2007). Artifacts and human concepts. In E. Margolis & S. Laurence (Eds.), *Creations of the mind: Theories of artifacts and their representation* (pp. 52–73). Oxford: Oxford University Press.
Uzgalis, W. L. (1988). The anti-essential Locke and natural kinds. *The Philosophical Quarterly, 38*, 330–339.
Verbeek, P.-P., & Vermaas, P. E. (2009). Technological artifacts. In J. K. B. Olsen, S. A. Pedersen, & V. F. Hendricks (Eds.), *A companion to the philosophy of technology* (pp. 165–171). Chichester: Wiley-Blackwell.
Wilson, R. A. (1999a). Realism, essence, and kind: Resuscitating species essentialism? In R. A. Wilson (Ed.), *Species: New interdisciplinary essays* (pp. 187–207). Cambridge, MA: MIT Press.
Wilson, R. A. (Ed.). (1999b). *Species: New interdisciplinary essays*. Cambridge, MA: MIT Press.
Wilson, R. A. (2005). *Genes and the agents of life: The individual in the fragile sciences – Biology*. Cambridge: Cambridge University Press.
Wilson, R. A., Barker, M. J., & Brigandt, I. (2007). When traditional essentialism fails: Biological natural kinds. *Philosophical Topics, 35*, 189–215.

Chapter 9
Ethnotechnology: A Manifesto

Beth Preston

Abstract The philosophical study of artifacts faces a serious problem going forward. There is a gap between work in the metaphysics of artifacts and work in other disciplines, such as psychology and anthropology. This gap primarily concerns the status of the category "artifact" itself and secondarily the status of the notion of artifact kinds. The existence of this gap raises questions as to whether work in the metaphysics of artifacts can be connected fruitfully with work on artifacts issuing from other disciplines, and if so, how. I argue that the best way to bridge this gap is a new interdisciplinary program I call ethnotechnology. I explain why ethnotechnology is needed and what it would involve.

Keywords Artifact kinds • Descriptive metaphysics • Ethnotechnology • Metaphysics of artifacts • Natural kinds

9.1 Introduction

An important interdisciplinary collection of essays about artifacts entitled *Creations of the Mind: Theories of Artifacts and Their Representation* (Margolis and Laurence 2007) appeared recently. In my review of this volume, I complained that the contributions by philosophers, almost all of which are devoted to the metaphysics of artifacts, have little apparent connection with the rest of the contributions.

B. Preston (✉)
Department of Philosophy, University of Georgia,
Athens, GA 30602-1627, USA
e-mail: epreston@uga.edu

M. Franssen et al. (eds.), *Artefact Kinds: Ontology and the Human-Made World*, Synthese Library 365, DOI 10.1007/978-3-319-00801-1_9,

> Especially troublesome…is the sense that the metaphysical preoccupations of the philosophers in Part I and the epistemological/cognitive preoccupations of the psychologists and assorted others in the rest of the volume are ships passing in the night with only the briefest of signals (about designers' intentions, would be my guess) flashing between them. (Preston 2008)[1]

Now there might be a lot of reasons for this apparent lack of connection. Perhaps the editors simply did not provide sufficient guidance for the reader as to what the connections were supposed to be. Perhaps a more varied selection of philosophers might have yielded better connectivity. No philosophers of technology are represented in this volume, for instance, although you would think their interest in artifacts is well attested. Or perhaps more contributors from anthropology might have helped to bridge the gap. If anybody has studied artifacts and material culture extensively, it is anthropologists and archaeologists, and they are not well represented in this volume.

But the more I have reflected on this shortcoming of an otherwise interesting and timely book, the more I have come to think that there is a much deeper and more general problem here, one that would not be remedied by a different set of contributors or more guidance from the editors. It is not a problem with this book. It is a problem about how the western tradition in metaphysics has been applied to the study of artifacts. If I am right about this problem, it bids fair to be a major stumbling block for the philosophical study of artifacts going forward – especially for the prospective role of philosophers in interdisciplinary research on artifacts. In the first section, I will explain the nature of the gap between the metaphysics of artifacts and work on artifacts in other disciplines. In the second section, I will explain how I think it should be interpreted by philosophers. And in the final section, I will suggest a way of bridging the gap in light of this interpretation.

To make the discussion manageable I will use the essays in the first two parts of *Creations of the Mind* as the material to be analyzed. The reader must bear in mind that the problem I am diagnosing is a general one, not a problem with this particular book or this particular collection of authors. This may be easier to do if the reader first of all bears in mind that most of these authors are reprising results and conclusions of previously published work, although usually adapting or extending them to fit the focus of this volume. So the gap is out there; this volume merely makes it visible in a way it has not been heretofore due to the diversity of disciplines and publishing venues involved.

9.2 Mind the Gap

Margolis and Laurence describe the contents of the four component parts of *Creations of the Mind* as follows (2007, pp. x–xi):

I. Metaphysics of artifacts
II. Concepts/categories of artifacts

---

[1] For a similar complaint by another reviewer, see Vermaas (2008).

III. Cognitive development of artifact concepts/categories
IV. Evolution of artifacts and artifact concepts/categories

Thus Part I deals with what artifacts really are, and Parts II–IV deal with how we conceive them, especially as this is evidenced in how we categorize them. The intended contrast, then, is between the metaphysics of artifacts and the epistemology of artifacts – understanding epistemology broadly to include cognition in general, not just knowledge strictly speaking. But how are the metaphysical issues supposed to be connected to the epistemological issues? We may get a sense of the difficulties that crop up here by comparing the results of the contributions in Parts I and II, respectively.

The central issues for the contributors in Part I, all of whom are philosophers, are most precisely stated by Richard Grandy.

> Why do we feel there are philosophical problems specific to artifacts? One reason, which is close to a common-sense philosophical intuition, is that artifacts – their existence and their features – depend on human interests. The second, more technical, is that it seems that if we believe something like the Kripke-Putnam story about kinds, artifacts lack essences and therefore aren't *real kinds*. (Grandy 2007, p. 21)

In short, there is a bothersome sense that artifacts are not real objects and that artifact kinds are not real kinds. So the philosophical problems specific to artifacts are ontological problems.

There is a history to this "intuition" that artifacts are ontologically suspect.[2] It began with Plato but was given its canonical form by Aristotle. In *The Republic* (476a–480a), Socrates says that the forms are "what is," whereas perceivable objects are "what is and is not." The rationale is that the forms alone are real because they are immutable, unitary, and universal, whereas perceivable objects change constantly, are multiplicitous (there are lots of blue things, and many shades of blue), and are relative to the observer (an object that looks blue to me may look green to you). This rationale rests on the unargued assumption that immutability, unity, and universality are criteria for reality. On this view, all ordinary objects – living beings and other natural objects as well as artifacts – are of a lower grade of reality than the forms. They are real insofar as they "participate" in the forms, but unreal insofar as they do not themselves have the reality-conferring characteristics of the forms.

Aristotle brings Plato's forms down to earth. The basic realities for Aristotle are primary substances, compounds of form and matter capable of separate and independent existence. Paradigmatic cases include individual trees or rabbits. Thus for Aristotle it is self-sufficiency – or autonomy, as Mary Louise Gill (1989, p. 213) puts it – that is assumed (again without argument, as far as I know) to be the criterion of reality. This allows Aristotle to rehabilitate living things, which independently maintain their own existence and reproduce their kind. But artifacts, although compounds of form and matter just like plants and animals, are unfortunately not independent in this way because they depend on their makers and users for their

[2] For another perspective on this history, one more focused on the last century in Anglo-American metaphysics; see Thomasson (2009).

origin, maintenance, and reproduction. So for Aristotle, artifacts are not substances, although they are analogous to substances in that they are compounds of form and matter (*Metaphysics* 1043a–b). Thus artifacts are again relegated to an ontologically second-class status, although on the basis of a different criterion.

Peter van Inwagen has advanced an influential modern version of Aristotle's view (1990). Because van Inwagen, unlike Aristotle, believes that matter is particulate, his view reaches its conclusion about artifacts by a variation on Aristotle's route (1990, p. 15). The elementary particles of matter (quarks, electrons, what have you) independently maintain themselves in existence and so are real on van Inwagen's view. And living things are real because they maintain their own unity and nature through changes in the particles that are their proper parts. Thus van Inwagen shares Aristotle's intuition that the criterion of reality is self-sufficiency, a characteristic that elementary particles and living things share. But on van Inwagen's view, artifacts do not merely have a lower grade of reality. Rather there are no artifacts (pp. 99–100). Because they do not independently maintain themselves in being through change, either as individuals or as kinds, as living things do, they are mere transient configurations of particles with no inherent unity and no proper parts. They are not things at all, strictly speaking, nor are rocks or rivers, for the same reason. Only living things escape van Inwagen's eliminativist move.

A number of other prominent contemporary metaphysicians have adopted similar eliminativist positions.[3] It is no wonder, then, that philosophers interested in ordinary objects like artifacts and nonliving natural objects have felt compelled to begin by arguing at length for the reality of their intended subject matter.[4] In the contemporary literature on artifacts, the Aristotelian worry about the dependent status of artifacts is typically formulated as the worry that artifacts are not real because they depend on human beliefs and intentions. The contributors to Part I of *Creations of the Mind* reprise the recent arguments.[5] John Searle, for instance, asks how there can be an "objective class of entities that exist only because we think they exist" (2007, p. 4). He accounts for these entities as "institutional facts" – realities constituted by language and human practices rather than by their sheer physical makeup. These social realities do "bottom out" in brute, physical facts, so there is a certain priority of physical facts over institutional facts, but this does not compromise the reality of the latter, on Searle's view. Artifacts are real in spite of their dependence on human thought. Crawford Elder and Amie Thomasson also address the question about the reality of artifacts in light of their alleged dependence on human intentions, but with particular attention to the reality of artifact

[3] Trenton Merricks (2000, 2001), like van Inwagen, eliminates all but living things at the macrolevel. Peter Unger (1979a, b) eliminates all macrolevel objects, including living things and even persons.

[4] Recent examples include Elder (2005), Baker (2007), and Thomasson (2007a).

[5] The only exception is Jerrold Levinson, who forgoes general questions about the reality of artifacts and artifact kinds to address the ontological status of works of conceptual art, where the intentions of the artist are allegedly more open ended than in the case of ordinary artifacts. I will therefore not discuss Levinson's contribution.

kinds. On Elder's view, artifact kinds are like natural kinds in being the result of a copying process which ensures the repeated cohesiveness of particular sets of characteristics. Moreover, the cultural copying process is just an extension of the one that results in natural biological kinds. So human intention is reduced to a detail of the implementation of cultural copying and no longer constitutes a reason to doubt that artifact kinds are as real as natural kinds. As Elder sums up his contribution: "What is written in the title of this chapter is 'artifacts'. What is written in the book of nature is *copied kinds*" (2007, p. 51). Thomasson, unlike Elder, agrees with Searle that artifacts and their kinds do depend on human intentions. But she agrees with both of them that artifacts and their kinds are nevertheless real, holding that "we should not assume that criteria suitable for determining whether or not there is a particular natural kind are suitable across the board, so we are not left in the position of either forcing artifactual kinds into the mold of natural kinds or denying their existence" (2007b, p. 73). In effect, then, Thomasson indicts the tradition back to Plato and Aristotle for begging the question about the reality of artifacts and other ordinary objects by making unargued and indefensible assumptions about the criteria for being real. Finally, Richard Grandy worries about mereology and its implications for the reality of artifacts. His question is how we can account for spatially discontinuous phases of artifacts (e.g., a disassembled bicycle, a scattered jigsaw puzzle) without falling prey to unrestricted constitution, which would allow any configuration of particles whatsoever to be an object. His answer is that even when separated, the parts of an artifact have a certain cohesiveness, since they can be – indeed, are intended to be – reassembled. So unlike random collections of particles, they still count as real, integral objects with proper parts even in their disassembled phases. Thus on Grandy's view, this feature of artifacts – that they can be and often are disassembled and reassembled – does not provide a good reason for doubting their metaphysical reality.

So much for the philosophers. Let us now turn to Part II, which is entitled "Concepts and Categories." The idea here is that we do not first have to settle the disputes about the reality of artifacts and artifact kinds to study how we conceptualize artifacts as opposed to natural objects and how we conceptualize, or categorize, them as falling into kinds. I will describe the contributions in Part II with special attention to the features that highlight their divergence from the metaphysics of artifacts elaborated in Part I.

Barbara Malt and Steven Sloman open their essay by stating that:

> If pressed…most cognitive psychologists would probably claim to be agnostic about whether or not there are objectively defined, metaphysically real groupings of artifacts…. [T]hey would suggest that what they mean by artifact 'kinds' are psychological kinds: groupings recognized by humans that might or might not correspond to the kinds that would be identified by the philosophers as metaphysically real. (Malt and Sloman 2007, p. 85)

This already seems like a polite way of saying: "We're just not talking about the same thing the philosophers in Part I are talking about." On the other hand, it might be merely a straightforward point about something that occurs with biological kinds as well. For example, in American English, the single biological order of the

Chelonians is categorized under two headings, "turtle" and "tortoise."[6] It is sometimes said that turtles are aquatic and tortoises are terrestrial, but there are exceptions like the box turtle, which is entirely terrestrial. So here it seems we just have some folk biological categories that do not track the scientific taxonomy very well and are ill defined to boot. But of course, there *is* no scientific taxonomy of artifacts as there is of living things and other natural objects. So the point as applied to artifacts is not so straightforward. Right off the bat, we have to wonder where exactly the philosophers are getting their artifact kinds and why these metaphysical kinds might diverge from recognized psychological kinds, especially if artifact kinds are – as most of the philosophers in Part I hold – mind-dependent.

But things get worse. The conclusion Malt and Sloman actually argue for is that "no coherent account of artifact categorization is possible, and 'categorization' is not a coherent field of inquiry" (p. 123). Their argument starts with the observation that the purpose of studying artifact categorization is to understand the cognitive underpinnings of daily life and activity. But, they continue, laboratory studies of artifact categorization often ask subjects to categorize things under constraints that occur nowhere in daily life, thus producing results that cannot be used for the intended purpose. Moreover, if we turn to examine the activities of daily life themselves, those activities that *do* seem to count as categorization are multifarious and produce quite different categorizations of artifacts depending on the situation, the task, the local language, and so on. So although there are clearly phenomena here to be studied, they do not amount to a single, integrated phenomenon we could call "*the* categorization of artifacts." Now we really have to wonder what the metaphysicians are up to. Because although Malt and Sloman are clearly trying to be diplomatic when they say they are studying psychological kinds as opposed to metaphysical kinds, their view suggests two possible interpretations of the ontological project of the philosophers. Either the purportedly "objectively defined, metaphysically real groupings" Malt and Sloman refer to are really just the parochial psychological groupings favored by philosophers for the purposes of arguing about whether or not there are real artifact kinds.[7] Or there are in fact objectively defined, metaphysically real groupings, but they bear even less relation to the multifarious and shifting psychological groupings of artifacts in daily life than scientific taxonomies of plants and animals bear to folk taxonomies.[8] In either case, though, the

[6] There is also a third category, "terrapin," which is not in wide use in the United States, although it is used more commonly in other English-speaking countries, I think.

[7] This interpretation sorts well with Thomas Reydon's view (Chap. 8, this volume) that the recent history of research into natural kinds in philosophy of science shows that kinds are in the first instance epistemologically defined groupings and that the metaphysics follows the epistemology rather than the other way around.

[8] This difficulty is illustrated in Crawford Elder's essay in Part I. He does have an objective definition of artifact kinds – they are created by biological or cultural copying processes. But his account of copied kinds has the result that neckties, high-heeled shoes, and the like are not copied kinds (2007, pp. 48–49). This is because the members of a copied kind must have a typical shape (literal or metaphorical), which is causally responsible for the successful performance that prompts the copying. Although neckties, for instance, do have a typical shape, Elder claims that this shape does

philosophers would appear to be caught up in projects, the results of which will not do much to illuminate our daily trafficking with artifacts or connect with the work of scholars in other disciplines who are aiming to illuminate it.

But things get worse. The next essay is by Dan Sperber, an anthropologist by training. His question is whether the Urcategorization on which this collection of essays is premised – the category "artifact" – is able to bear the theoretical weight placed upon it. His conclusion is that it cannot, at least if it is defined, as it usually is, in terms of paradigmatic cases like hammers or bracelets. Sperber's argument rests largely on an analysis of the nonparadigmatic case of domesticated plants and animals. He employs Ruth Millikan's notion of proper function for this purpose – a notion which, it should be noted, Millikan explicitly designed to apply to natural and cultural objects alike. But Sperber argues that nature and culture are blurred beyond recognition in the case of domesticates, since the organisms involved carry out their cultural functions in virtue of carrying out their biological functions. Moreover, these biological functions are themselves often adaptations to cultural conditions, for example, seedlessness in grapes or wooliness in sheep. He also observes that from the beginnings of agriculture until the industrial revolution, these biological artifacts were more common in daily life than the currently more paradigmatic hammers and bracelets.

> Here I have tried to cast doubt on the idea that a theoretically useful notion of artifact can be built around its usual prototypes: bracelets, jars, hammers and other inert objects, or that it can be defined in a more systematic way. There is a continuum of cases between public productions that are well characterized by a specific purpose and others where purpose is unclear. There is also a continuum of cases between public productions that are wholly designed by humans, and others where humans exploit, with little or no modification, a pre-existing structure. …There is no good reason why a naturalistic social science should treat separately, or even give pride of place to, cultural productions that are both more clearly intended for a purpose and more thoroughly designed by humans, that is, to prototypical artifacts. (Sperber 2007, p. 137)

If Sperber's conclusion is right, the preoccupations of the metaphysicians rest on a local categorization which can be made conceptually rigorous only by ignoring importantly ambiguous phenomena like domestication. But more importantly for our purposes, if Sperber's argument is right and nature and culture are a seamless fabric, ontology ought to reveal a seamless fabric as well. This suggests that Aristotle set metaphysics off on a path that was bound to diverge radically from the "naturalistic social science" represented by the extra-philosophical contributors to *Creations of the Mind* insofar as he did treat artifacts separately from living things. And it

---

not cause the performance of generating social acceptability for the wearer which is responsible for the proliferation of neckties. Rather it is the behavior of wearing a necktie (which also has a typical shape, if not a literal one) which generates the social acceptability. So counterintuitively – at least from the perspective of the folk – neckties and the like are *not* copied kinds and thus not artifact kinds, whereas wearings of neckties and similar behaviors *are* copied kinds, and thus presumably artifact kinds. Whether or not Elder's analysis here is correct, it illustrates nicely the insouciance with which even metaphysicians dedicated to the philosophical rehabilitation of ordinary objects may be disposed to treat ordinary intuitions about artifacts and their kinds.

suggests that philosophers might do well to reconsider whether treating artifacts separately in Aristotle's wake, as they tend to do, really has the theoretical usefulness they tend to suppose.

To be fair to the philosophers, some of them do allude to the difficulties attendant upon a rigorous definition of "artifact." This is particularly true of Richard Grandy, who discusses at length some of the reasons for believing there is a continuity rather than a sharp divide between artifacts and natural objects and who concludes explicitly that "it doesn't seem that there should be a metaphysical distinction" (Grandy 2007, p. 24). Similarly, Crawford Elder's conclusion that there are only copied kinds implies that natural kinds and artifact kinds form a seamless fabric. But neither Grandy nor Elder goes on to question, as Sperber does, whether the category "artifact" is of any real use for theoretical purposes. Indeed, if Sperber is right, it is a hindrance in our attempts as naturalistic social scientists to understand the role played by the things we make and/or use in everyday life. He might, then, be best taken as recommending that philosophers consider the possibility that this category is inhibiting their efforts in ontology as well, and in addition compromising possibly fruitful connections between philosophy and social science. Sperber's challenge, in short, is not so much that philosophers have failed to acknowledge the continuity between nature and culture – although some certainly have – but that even those who do acknowledge it have not drawn the proper conclusions with regard to philosophical theory and practice.

The three remaining contributors to Part II can best be interpreted as providing support for Sperber's view in the form of further elaboration of the sort of problems he raises. Paul Bloom casts doubt on the idea that there is a sharp distinction between artifacts and natural objects by presenting evidence that we think of many things as simultaneously artifactual and natural. His main example is water.

> There are many cues that it [water] is a natural kind. It falls from the sky, after all, and is found in oceans, rivers, and lakes. But there are also good reasons to take it as an artifact kind. It comes from bottles, cans, taps, hoses, and coolers; it is filtered, processed carbonated, purified, and chlorinated; it is advertised on television and sold in stores. The sensible conclusion for children to draw from these facts is that water is both a natural and an artifact kind. (Bloom 2007, pp. 155–156)

Bloom calls the concepts children form under such conditions "hybrid concepts" and asserts there are many other cases of such concepts. This supports Sperber's view that nature and culture are a seamless fabric and in addition suggests that our ordinary concepts in fact reflect this. Hilary Kornblith, the lone philosopher included in Part II, in effect concludes that our linguistic practices reflect this seamlessness as well. Kornblith does think there are metaphysical differences between artifact kinds and natural kinds. But, he argues, these differences have no influence on our linguistic practices, at least as far as the theory of reference is concerned. The currently favored theory is that names are introduced into the language through a "baptism" event which connects the name with a certain kind of thing, and this usage then diffuses throughout the linguistic community. This account works just as well in the case of artifacts as in the case of natural kinds, according to Kornblith. So although he may resist the idea that nature and culture are a seamless fabric, the linguistic evidence he marshals actually points to the opposite conclusion.

The last contribution in Part II casts doubt on the category of "artifact" in a different way. It is a long and detailed review by Bradford Mahon and Alfonso Caramazza of the current evidence for competing theories about neural representation and organization of concepts. They end with the comment that these theories are at present underdetermined by the data and they plead for more subjects to be identified and tested. So their specific conclusions are at best hard to evaluate. But it is actually an assumption they make going in that is significant for our purposes.

> The position for which we will argue assumes that the first-order constraint on the organization of conceptual and perceptual processes is object domain, and that the candidate domains are those that could have been evolutionarily salient in our phylogenetic history: living animate, living inanimate, con-specifics, and possibly tools. (Mahon and Caramazza 2007, p. 158)

Mahon and Caramazza gloss "tool" as "manipulable artifact" and "living inanimate" as "fruit/vegetables" (p. 155). But if these really are the most basic categories of objects into which we parse the world, it is notable that there *is* no category of "artifact" as such. Indeed, this seems like a strange list for what it leaves out – all the non-manipulable artifacts (shelters, clothing, monuments, boats, and so on) as well as all the nonliving inanimate things (rocks, dead stuff, water, and so on). Presumably these omissions reflect evolutionarily less salient domains of objects – domains that would be the subject of categorization only at a further, nonbasic level. But that still leaves us with the worry that there is no category of artifact as such on this view, not because there is no divide between nature and culture, but because artifacts as such do not seem epistemically salient from an evolutionary perspective on human cognition.

In summary, a gap has opened up between the epistemologists and the metaphysicians. To begin with, the epistemologists are skeptical that the category "artifact" carves the world at a joint, whereas the metaphysicians take this category pretty much for granted. Although the epistemologists do not all explicitly reject the nature-artifact distinction on theoretical grounds, their results and/or auxiliary assumptions pretty uniformly cast doubt on its integrity and usefulness. Second, some of the epistemologists are equally skeptical about the artifact kinds taken for granted by the metaphysicians. We already know there is no scientific taxonomy of artifacts, and psychological studies of categorization do not support the existence of a folk taxonomy of artifacts of the sort we think we have for natural objects, and in particular for animals and plants. In short, the two most basic assumptions employed by the metaphysicians – assumptions that are fundamental background for the questions they ask about reality – are not shared by the epistemologists.

9.3 Whose Metaphysics? Which Ontology?

How are we to interpret this gap? An obvious option would be to say that things were ever thus. Metaphysicians typically discover that what is truly real is not what ordinary people think is real and/or that the way ordinary people carve up the world is not the way the world should be carved up if the carving is to respect actual joints.

In the passages we referred to above, for instance, Socrates says that the "lovers of sights and sounds" do not credit the existence of the forms at all and instead believe that the things they perceive are truly real. And I don't imagine the ordinary person in the street would be any more likely to credit Peter van Inwagen's view that there are no tables or rocks. In other words, metaphysicians are typically engaged in *revisionary* rather than *descriptive* metaphysics, in Strawson's (1959/1963) sense.[9] So there is nothing to wonder at or even be concerned about if the results in metaphysics diverge from the ordinary beliefs and practices investigated by the "naturalistic social scientists" who make up the majority of the epistemology wing here.

But perhaps this is not the right interpretation of Part I. Perhaps at least some of the metaphysicians are instead engaged in descriptive metaphysics. In this case, the gap dividing them from the epistemologists is much more worrying, because unlike the revisionary metaphysician, the descriptive metaphysician does not have the luxury of simply ignoring what ordinary people think. Descriptive metaphysics is supposed to uncover the basic ontological categories to which we as human beings are actually committed (Strawson 1959/1963, p. xiv). So if the results obtained by the epistemologists – who are actively studying the thinking and practices of ordinary human beings, after all – indicate that we (the folk) do not, in fact, take "artifact" to be a distinct, basic category of beings or that our cognitive processes do not, in fact, require or construct stable and distinct artifact kinds, then something has gone badly wrong with the metaphysicians' descriptive enterprise. But this interpretation leaves us wondering: how *could* the metaphysicians have gone so badly wrong?

Here is a hypothesis that goes some way towards answering this last question while at the same time revealing further complexities. Perhaps what the descriptive metaphysicians are describing are local metaphysical commitments that are not necessarily widely shared among the folk in general. Strawson himself, in good Kantian fashion, declared that there are some very basic ontological commitments of an ahistorical, acultural nature central to the thinking of human beings as such. And he says descriptive metaphysics is "primarily" concerned with these (1959/1963, p. xiv). But he explicitly leaves open the possibility of more fine-grained descriptions that would capture less basic ontological commitments that do vary from time to time and culture to culture.[10] So we may conjecture that when the descriptive metaphysicians speak of artifacts and artifact kinds, they may be reflecting the ontological commitments of their own academic folk – commitments that are not necessarily widely shared even among other academic folk, let alone among ordinary, nonacademic folk in western culture, let alone among human folk as such. And because the naturalistic social scientists are casting their nets more widely among the folk, the gap we have been examining has opened up between their enterprises and those of the philosophers.

---

[9] "Descriptive metaphysics is content to describe the actual structure of our thought about the world, revisionary metaphysics is concerned to produce a better structure" (Strawson 1959/1963, p. xiii).

[10] "Certainly concepts do change, and not only, though mainly, on the specialist periphery; and even specialist changes react on ordinary thinking" (Strawson 1959/1963, p. xiv).

These interpretations are not mutually exclusive, of course. In some cases, the gap may be due to a revisionary metaphysics, while in others it may be due to a descriptive metaphysics with a limited range. Either way, though, it seems that more reflection on the scope and limits of existing metaphysical projects is called for, as well as more reflection on the possibility of a descriptive metaphysics with a greater ability to connect with the work of the naturalistic social scientists – a descriptive metaphysics that could close the gap.

9.4 Ethnotechnology: The Manifesto

To begin this reflection, let us examine the options available to the metaphysician. One possibility would be to simply opt for revisionary metaphysics. We may call this the revisionary option. A second possibility would be to opt for descriptive metaphysics, but not claim to be describing the ontological commitments of any wider tradition than that of the philosopher folk themselves. We may call this the traditional option. The third and final possibility would be to opt for descriptive metaphysics but to make a concerted effort to uncover and analyze the ontological commitments of a wide variety of folk, especially as these commitments bear on things the folk make and/or use. This is the naturalistic social science option. We will discuss the first two options and their limitations briefly, before moving to a more extended discussion and exhortation in favor of taking up the naturalistic social science option.

Let us start with the revisionary option. You might think the revisionary metaphysician could just continue on without worrying about the activities of the naturalistic social scientist or the folk she studies. But there is at least one constraint on this freedom. Even *revisionary* metaphysicians should be concerned about theoretically based arguments of the sort Sperber advances. For even revisionary metaphysicians are trying to carve the world at its joints – indeed, they are trying to do a better job of it than ordinary people do – and if Sperber is right, "artifact" is not a joint. The fact that Grandy – a philosopher classified with the metaphysicians – echoes Sperber's views is further evidence that the metaphysicians, whether revisionary or descriptive, are neglecting an important, prior question when it comes to the metaphysics of artifacts. Sperber and Grandy are not trying to describe how ordinary people think about the world, but rather to describe the world itself, to take account of the phenomena not as seen by the ordinary person necessarily, but rather as theorized by scientists and philosophers. From this point of view, Sperber's essay is in the wrong section of the book. He belongs with the metaphysicians, and perhaps even with the revisionary wing, since he is clearly more concerned about what there is than about what the folk think there is. Except for this rather important question, though, revisionary metaphysicians may safely ignore the folk and the naturalistic social scientists who study them. If the folk conceptualize some things as ambiguously natural and artificial or if they do not categorize them reliably at all, that is not a problem for the revisionary metaphysician, who after all is in the business of tidying up exactly this sort of messiness.

The limitation of revisionary metaphysics, though, also stems from the lack of external constraint. It makes for an insular enterprise with little in the way of connecting bridges even to other branches of philosophy, let alone other academic disciplines or the folk at large. This does not diminish the importance of revisionary metaphysics in principle, since the revisionary metaphysician may, of course, be right about what there is. Similarly, the theoretical physicist may be right, no matter how bizarrely different her conception of the universe from that of the ordinary person. On the other hand, it does seem to me that there is a problem in practice. For what is the point of being right if this does not have any effect on what the rest of the world thinks or does? As Strawson points out, one of the ways concepts change is through influence from the "specialist periphery" (see footnote 10 above). But in the case of revisionary metaphysics, a lack of connecting bridges makes for no communication going out from the specialist periphery, not just for no influence coming in from the nonspecialist masses. Things do not have to be this way, though. Theoretical physics has done a better job of this, for instance. Most educated people have heard about string theory; how many have heard about mereological nihilism? An even more pertinent example is Buddhist metaphysics, which typically resembles nothing so much as the radical eliminativism of Peter Unger – i.e., not just no tables and rocks, but no living beings or persons either. But this startling metaphysics is firmly connected to a set of practices that is in principle available to everyone, and that is aimed at the alleviation of suffering and the promotion of compassion for all sentient beings on the part of the practitioner.[11] So in the case of these eastern revisionary metaphysicians, the connecting bridges have been carefully maintained. There is no reason in principle why western revisionary metaphysicians could not follow the lead of the theoretical physicists or the Buddhist metaphysicians if they wanted to. And of course if they do not want to, that is their prerogative.

Be that as it may, our more important business here is with descriptive metaphysics. So let us consider the traditional option. This involves proceeding as metaphysicians interested in artifacts have in fact done, working within the philosophical tradition they have inherited using the methods it makes available. This option is well represented by the philosophers in *Creations of the Mind*. Some of them are concerned to counter the depredations of the revisionary metaphysicians who insist there are no artifacts. Others are using the resources of philosophy of language to examine questions about artifact kinds. Still others are concerned with standard mereological puzzles as they apply to artifacts, and so on. As with revisionary metaphysics, this work is, in principle, very valuable, especially insofar as it serves to correct whatever errors the tradition may have accumulated.[12] On the other hand, it keeps the metaphysics of artifacts within the ambit of some traditional assumptions, including the following: that "artifact" is a basic category, that there is

[11] The connections are very complex, though, and do not always result in the lay practitioner fully understanding or accepting the metaphysical theory that backs the practices. For an influential and sophisticated account of these connections in the case of *Theravada* Buddhism, see Melford Spiro (1982).

[12] Thanks to Amie Thomasson (personal communication, 2010) for insisting on this point.

a significant divide between nature and culture, and that there are artifact kinds just as there are natural kinds. And what I have been concerned to argue in this chapter is that proceeding this way bids fair to consolidate the disciplinary isolation these metaphysicians already enjoy, since academics in other disciplines increasingly question precisely these assumptions on the basis of their investigations of how ordinary people think about artifacts. In short, these assumptions seem to reflect the ontological commitments of the philosopher folk and not those of the folk at large. So while the metaphysicians really are carefully describing the phenomena as they see them, what phenomena they see and/or attend to is, not surprisingly, influenced by the very special subculture to which they belong and its attendant history. The end result is a metaphysics of artifacts that is descriptive on its native turf but is in effect revisionary if applied to the folk at large.

One limitation of this kind of descriptive metaphysics, then, is a kind of insularity similar to that engendered by explicitly revisionary metaphysics. It is evidenced by the gap we have been discussing between the philosophical and nonphilosophical contributors to *Creations of the Mind*. This gap limits the potential of the descriptive metaphysician to contribute in a robust way to what is in principle an interdisciplinary area of research concerned with artifacts. As in the case of revisionary metaphysics, the traditional option results in a lack of influence in either direction between the descriptive metaphysicians and the social scientists. Another related limitation concerns the potential of traditional descriptive metaphysics to contribute to what we might call a general philosophy of artifacts – a philosophical investigation into the role in people's lives and activities of the things they make and/or use. For that role is arguably undergirt by the ontological commitments of ordinary people, not the specialized commitments of the metaphysical tradition. So to the extent that the metaphysicians are limiting their descriptive base to the ontological commitments of the philosophical tradition, they are unlikely to produce results that would be useful to a more general philosophical investigation into the role of artifacts in human existence. Thus the traditional option limits descriptive metaphysics with regard to its participation in both interdisciplinary projects and more general philosophical ones concerning artifacts.

This brings us to the third option. Instead of sticking with the traditional concerns and methods of philosophical metaphysics, an adventurous descriptive metaphysician might strike out in a new direction and launch a program to discover and articulate the folk ontologies concerned with artifacts. This would be very ambitious, since it is clear that there is not just one such folk ontology, but many. They may have some common core but for the rest may vary substantially by culture and even subculture. For example, in industrialized cultures, there are usually specialized groups of people who deal with the design and engineering of artifacts, especially the more complex types thereof. And the design and engineering folk may well have a different way of categorizing or conceptualizing the artifacts they deal with on a professional basis than the nonspecialist end users of those artifacts do.[13]

[13] See the contributions in Part III of this volume for some interesting suggestions along these lines.

Such a program would also require empirical studies of the relevant folk, and since philosophers are not usually well equipped to do this on their own, interdisciplinary arrangements with social scientists would be required. So this third option is properly called the naturalistic social science option. But what should we call the program itself? I propose calling it *ethnotechnology*. In recent years, a number of similar programs have cropped up in anthropology and sociology, e.g., ethnomusicology, ethnobiology, and ethnomedicine. In the most general terms, these are the study of how people make and use music, how they relate to and use the biological world, how they discover, produce, and use medicines, and so on. So ethnotechnology would be the study of how people make and/or use artifacts.[14] But this proposal leaves us with two important questions. First, what resources are currently available for getting such a program off the ground? Second, why *should* a descriptive metaphysician strike out in this new direction? We will take these questions up in turn.

So how are philosophers to study these folk ontologies? In particular, what empirical methodologies would be most appropriate? One possibility is experimental philosophy.[15] This relatively recent movement engages philosophers not just in using empirical data (which they have pretty much always done) but in generating it as well. Experimental philosophers use the methods of experimental psychology to get at the intuitions of the folk about various concepts of interest to the philosophers – intention, free will, moral responsibility, and the like – and to establish the psychological underpinnings of these intuitions. Typically, this involves presenting subjects with thought experiments or scenarios and asking questions designed to elicit their intuitions about them. This is a welcome corrective to the all too common tradition of presenting the reflections of a lone philosopher snuggled in the depths of his armchair as what "we" think or as "our" intuitions. So experimental philosophy does have the virtue of aiming at what the folk really think and how what they think bears on some important philosophical issues. But there is also a problem with this experimental methodology for our purposes. As Malt and Sloman complain, experimental methodologies in psychology bring subjects into a highly constrained environment and set them to doing highly constrained tasks and/or answering very explicit questions. In the process, what is purportedly to be explained – the activities of daily life – is typically bypassed or distorted (2007, pp. 116–119). We may add to this that the questions about artifacts and their role in human existence are not just about what people *think*, but about what people *do*. And we have to reckon with the very real possibility that in many cases what people do is not reflected very well, if at all, in what

[14] The term "ethnotechnology" has sometimes been used as a replacement for the outdated and undesirable "primitive technology" to mean the study of traditional technologies only. See http://deletionpedia.dbatley.com/w/index.php?title=Ethnotechnology_(deleted_13_Feb_2008_at_18:26), for instance. I intend it to be used generally to cover the study of people making and/or using things, regardless of the level of technical sophistication involved.

[15] The experimental philosophers have their own manifesto (Knobe and Nichols 2008). Or see the experimental philosophy homepage: http://experimentalphilosophy.org/Experimental-Philosophy.html

they think. This is all the more likely to be a difficulty with regard to our transactions with artifacts, which, unlike ethical dilemmas or questions about free will, are too routine and unproblematic to rise to the level of consciousness. In light of these considerations, then, what we need is a methodology that is empirical but is not wedded to the laboratory or narrowly focused on the psychology of the folk.

But we are in luck. Just such a methodology already exists and is in routine use in anthropology, archaeology, and related fields for describing and analyzing the transactions of the folk with the natural world. This program is called ethnobiology.[16] It uses the fieldwork methods standard in anthropology, including interviews with local informants and participant observation. Additional methods include archaeological techniques for obtaining data for cultures that are no longer extant and statistical techniques for analysis of data. Early work in this area was largely focused on folk classifications of plants and animals and on comparing these classification systems from cultures worldwide with western scientific taxonomy.[17] But as the field has grown, the focus has expanded to include study of the uses people make of plants and animals and the knowledge systems that comprehend both the classification and the use aspects of the relationships between people and their local biota. Importantly for our purposes, this has recently led to a burgeoning interest in domestication and medicine, among other things.

Because a huge literature on folk classification already exists in ethnobiology, complete with multiple methodologies for assessing these classifications in situ, ethnobiology would provide especially rich resources for philosophers interested in artifact kinds. More importantly, perhaps, since it is already an inherently interdisciplinary enterprise, integrating a few philosophers should not present any serious difficulties.[18] But how does the study of artifacts fit with the existing focus on our relations with the natural environment in ethnobiology? First of all, if nature and culture are indeed a seamless fabric – as Sperber argues and as some philosophers already agree – the study of artifacts is a natural extension of the study of plants and animals. Second, ethnobiologists are in fact already studying artifacts – domesticates, for instance. They also already study the processing and use of plants and animals in foodways and medicine, and what is in your cooking pot or medicine bottle is clearly artifactual. Perhaps more pointedly, your cooking pot and medicine bottle are essential to the processing and use of whatever is in them, so the ethnobiologists can hardly fail to be taking artifacts into their purview, no matter what. Philosophers with an interest in artifacts would only be asking them to do this more explicitly.

[16] A good introduction to ethnobiology can be obtained from the website of the Society of Ethnobiology (http://www.ethnobiology.org/). There is also a related research area called ethnoecology. It takes a more holistic approach, focusing on the relationships between people and the ecosystems in which they live. This approach would also be a fruitful one to pursue, but to keep things simpler and within space limitations, I will not discuss it here.

[17] For a classic work in this vein, see Brent Berlin's *Ethnobiological Classification* (1992).

[18] Indeed, some philosophers of science are already involved. See the essays by John Dupré and David Hull in Medin and Atran (1999), for instance.

But why should a descriptive metaphysician choose this challenging path? The short answer is: to overcome the limitations of the revisionary and traditional descriptive options we discussed above – i.e., to unlock the potential of philosophical metaphysics to contribute fully to interdisciplinary investigations of artifacts and to a more general philosophy of artifacts. But this just raises further questions calling for a longer answer. Why worry about overcoming these limitations? Why is the ethnotechnology program important enough to warrant such a radical deviation from the well-trodden paths of traditional metaphysics?

In the history of western philosophy, there has never been an area of research devoted to artifacts as such. As we have seen, even western metaphysics considered artifacts only with an eye to fitting them into some grand, overarching ontological scheme driven by concerns very foreign to our everyday life with things. Some types of artifacts have, indeed, come in for serious attention – in particular, works of art and, more recently, the sophisticated modern artifacts favored by the new area of philosophy of technology. But there has been no philosophy of artifacts. Nor has there been any other discipline or science devoted entirely to artifacts as such. Some types of artifacts have been studied in more practically oriented fields like engineering. But it is only in the last century or so that fields like archaeology, anthropology, and design studies have sprung up and have devoted a major share of their resources to a general study of artifacts and their role in human life. It is a good question why artifacts have not received very much attention until comparatively recently. But we will let that question go in order to concentrate on an argument for giving them our full attention now.

For the first 100,000 or so years of human history, we made our living by hunting and gathering. Even after the beginnings of agriculture about 10,000 years ago, most of us still made our living that way. Except in a few, particularly resource-rich areas, hunting and gathering is necessarily a nomadic way of life. Nomadic peoples do not either need or want a lot of artifacts. Especially not heavy or breakable things that are difficult to carry around on your back. (No draught animals, remember.) Typically, they also go in for artifacts that can be easily made from locally available materials. That way you can abandon your old ones and make new ones wherever you end up. By all accounts, this was (and is) a completely sustainable way of life as far as use of the planet's resources goes.[19] It is with agriculture and the settled way of life it made possible that we began to accumulate artifacts,[20] not just more of them, but types of artifacts that had not existed before, like pottery and textiles. And for the last 10,000 years, the proliferation of artifacts has continued to accelerate. We have now reached a point at which our way of life may not be sustainable in large part because the vast array of consumer goods it involves uses up too many resources which are either not renewable or cannot be renewed fast enough to keep

[19] The sustainability also depends to some extent on the limitation of population growth. Hunter-gatherers typically space their children out at four- or five-year intervals. This is because you do not want to have more than one child at a time who has to be carried anymore than you want a lot of material possessions that have to be carried.

[20] Populations started to grow much faster at this period as well.

up with consumer demand. Connected with this problem of sustainability is a problem of distributive justice, since the proliferation of artifacts has also made it possible for some people – or some cultures – to accumulate a lot more of them than others.

The proliferation of artifacts and the problems of sustainability and justice it brings with it is something Plato was already concerned about in the *Republic*. This comes out especially clearly in his comparison of the healthy "city of pigs" with the luxurious "city with a fever" (*Republic*, 370b–373a). He has Glaucon insist that the city of pigs, in which only basic survival and comfort needs are met, is not the kind of society Socrates should consider as the basis for the ideal city, but rather the feverish city which generates and meets a plethora of other nonbasic needs as well. Now Plato does not talk about this contrast in terms of artifacts: he talks about luxury or wealth. But wealth and luxury are overwhelmingly constituted by artifacts of one sort or another. For example, the inhabitants of the city of pigs wear clothes only for warmth in the winter and go naked in the heat of the summer, eat their meals off leaves or reed mats, recline on heaps of leaves and branches, and have a vegetarian diet. The city with a fever, on the other hand, has luxury clothing, proper tableware and furniture, and meat dishes and pastries in abundance. In short, the healthy city of pigs is a society oriented to minimal production and use of artifacts. In contrast, the city with a fever escalates the production and use of different kinds of artifacts embodying the luxuries it requires. And of course, what were luxuries in Plato's time would represent only a minimally acceptable living standard today – at least in the "developed" countries of the west.

Socrates says as much as that the city with a fever will soon run out of local resources and have to resort to war to take resources from other cities. The city of pigs, then, is healthy in large part because it is what we would now call sustainable – it lives within its resources. On the other hand, the luxurious city is caught up in a fevered drive to accumulate more and more artifacts and other forms of wealth at the expense of its resource base. So clearly Socrates regards the feverish city as a problematic kind of human society, and just as clearly the main root of its problems is all the stuff it wants and needs. On the other hand, Socrates does not long resist Glaucon's claim that the city with a fever and not the city of pigs is the proper basis for their discussion. He gives two reasons for giving in. First, Socrates says, the city with a fever is a better laboratory for examining the origin of justice and injustice. The idea here, apparently, is that the healthy city of pigs would be automatically just, and it would then be hard to say what makes it just since there would be no injustice for comparison. Thus Socrates, too, connects the question of sustainability with the question of justice. Second, Socrates says that they may as well consider the city with a fever because that is what people will want anyway, no matter what he tells them about healthy and unhealthy ways of life.

> All right, I understand. It isn't merely the origin of a city that we're considering, it seems, but the origin of a *luxurious* city. And that may not be a bad idea, for by examining it, we might very well see how justice and injustice grow up in cities. Yet the true city, in my opinion, is the one we've described, the healthy one, as it were. But let's study a city with a fever, if that's what you want. There's nothing to stop us. The things I mentioned earlier and

> the way of life I described won't satisfy some people, it seems, but couches, tables and other furniture will have to be added, and, of course, all sorts of delicacies, perfumed oils, incense, prostitutes, and pastries. We mustn't provide them only with the necessities we mentioned at first, such as houses, clothes, and shoes, but painting and embroidery must be begun, and gold, ivory, and the like acquired. (Plato 1992, pp. 47–48; *Republic*, 372e–373a)

In this passage, Socrates suggests the proliferation of artifacts cannot be stopped or reversed, so he sees no choice but to accommodate it in his vision of the ideal society.

But we may now have reached a point where accommodation is no longer an option. Plato only had to worry about the sustainability and justice of a single city state among others. We have to worry about the sustainability of human existence on this planet with (one would hope) some measure of distributive justice for all. Plato already points us in the direction of artifacts and the relationships people have to them as lying at the heart of these problems. But perhaps because he decides to accommodate the proliferation of artifacts, he is not prompted to mount an investigation into artifacts as such, and their actual and possible roles in human existence.

Perhaps if he had, and if a philosophy of artifacts had consequently become a central element of our philosophical and intellectual tradition, we would have a better grip on the questions about sustainability and distributive justice that plague us today. But what this means is that we now have pressing reasons to do what Plato did not do and make a philosophy of artifacts central to the philosophical enterprise. In turn, a nontraditional descriptive metaphysics is required for addressing the wider epistemological, political, social, and ethical questions about artifacts such a philosophy of artifacts would include. For answers to these questions depend on understanding the folk ontologies of artifacts that undergird how people not only think about but actively deal with artifacts in their daily lives. So we may hope that some philosophers will go for the naturalistic social science option in metaphysics and engage in the interdisciplinary work necessary for such an account of artifacts and their role in human life.

References

Baker, L. R. (2007). *The metaphysics of everyday life: An essay in practical realism*. Cambridge: Cambridge University Press.

Berlin, B. (1992). *Ethnobiological classification*. Princeton: Princeton University Press.

Bloom, P. (2007). Water as an artifact kind. In E. Margolis & S. Laurence (Eds.), *Creations of the mind: Theories of artifacts and their representation* (pp. 150–156). Oxford/New York: Oxford University Press.

Elder, C. L. (2005). *Real natures and familiar objects*. Cambridge, MA: MIT Press.

Elder, C. L. (2007). The place of artifacts in ontology. In E. Margolis & S. Laurence (Eds.), *Creations of the mind: Theories of artifacts and their representation* (pp. 33–51). Oxford/New York: Oxford University Press.

Gill, M. L. (1989). *Aristotle on substance*. Princeton: Princeton University Press.

Grandy, R. E. (2007). Artifacts: Parts and principles. In E. Margolis & S. Laurence (Eds.), *Creations of the mind: Theories of artifacts and their representation* (pp. 18–32). Oxford/New York: Oxford University Press.

Knobe, J., & Nichols, S. (2008). An experimental philosophy manifesto. In J. Knobe & S. Nichols (Eds.), *Experimental philosophy* (pp. 3–16). New York: Oxford University Press.

Mahon, B. Z., & Caramazza, A. (2007). The organization and representation of conceptual knowledge in the brain: Living kinds and artifacts. In E. Margolis & S. Laurence (Eds.), *Creations of the mind: Theories of artifacts and their representation* (pp. 157–187). Oxford/New York: Oxford University Press.

Malt, B. C., & Sloman, S. A. (2007). Artifact categorization: The good, the bad, and the ugly. In E. Margolis & S. Laurence (Eds.), *Creations of the mind: Theories of artifacts and their representation* (pp. 85–123). Oxford/New York: Oxford University Press.

Margolis, E., & Laurence, S. (Eds.). (2007). *Creations of the mind: Theories of artifacts and their representation*. Oxford/New York: Oxford University Press.

Medin, D. L., & Atran, S. (Eds.). (1999). *Folkbiology*. Cambridge, MA/London: MIT Press (Bradford Books).

Merricks, T. (2000). No statues. *Australasian Journal of Philosophy, 78*(1), 47–52.

Merricks, T. (2001). *Objects and persons*. Oxford: Clarendon Press.

Plato. (1992). *Republic* (G. M. A. Grube, Trans.). Indianapolis: Hackett Publishing Company.

Preston, B. (2008). Review of *Creations of the mind: Theories of artifacts and their representation* (E. Margolis, & S. Laurence, Eds.). Oxford/New York: Oxford University Press, 2007, *Notre Dame Philosophical Reviews*, May 14, 2008. http://ndpr.nd.edu/review.cfm?id=13083

Searle, J. R. (2007). Social ontology. In E. Margolis & S. Laurence (Eds.), *Creations of the mind: Theories of artifacts and their representation* (pp. 3–17). Oxford/New York: Oxford University Press.

Sperber, D. (2007). Seedless grapes: Nature and culture. In E. Margolis & S. Laurence (Eds.), *Creations of the mind: Theories of artifacts and their representation* (pp. 124–137). Oxford/New York: Oxford University Press.

Spiro, M. E. (1982). *Buddhism and society: A great tradition and its Burmese vicissitudes* (2nd, expanded ed.). Berkeley/Los Angeles/London: University of California Press.

Strawson, P. F. (1959/1963). *Individuals*. Garden City: Doubleday & Company, Inc. (Anchor Books).

Thomasson, A. L. (2007a). *Ordinary objects*. Oxford: Oxford University Press.

Thomasson, A. L. (2007b). Artifacts and human concepts. In E. Margolis & S. Laurence (Eds.), *Creations of the mind: Theories of artifacts and their representation* (pp. 52–73). Oxford/New York: Oxford University Press.

Thomasson, A. L. (2009). Artifacts in metaphysics. In A. Meijers (Volume Ed.) and D. M. Gabbay, P. Thagard, & J. Woods (General Eds.), *Handbook of the philosophy of science. Vol. 9: Philosophy of technology and engineering sciences* (pp. 191–212). Amsterdam/Oxford/Burlington: Elsevier.

Unger, P. (1979a). There are no ordinary things. *Synthese, 41*, 117–154.

Unger, P. (1979b). Why there are no people. In P. A. French, T. E. Uehling, & H. K. Wettstein (Eds.), *Studies in metaphysics* (*Midwest Studies in Philosophy* 4, pp. 177–222). Minneapolis: University of Minnesota Press.

van Inwagen, P. (1990). *Material beings*. Ithaca/London: Cornell University Press.

Vermaas, P. E. (2008). Review of *Creations of the mind: Theories of artifacts and their representation* (E. Margolis, & S. Laurence, Eds.). Oxford/New York: Oxford University Press, 2007, *Philosophy of Science, 75*(October 2008), 473–477.

Part III
Artefact Kinds and Engineering Practice

Chapter 10
On What Is Made: Instruments, Products and Natural Kinds of Artefacts

Wybo Houkes and Pieter E. Vermaas

Abstract Debates in the metaphysics of artefacts typically start from the observation that technical artefacts result from intentional production and then focus immediately on the issue whether this 'mind-dependence' undermines claims that artefacts exist or come in natural or real kinds. We aim to add sophistication to debates on the latter issue by approaching it through an analysis of contemporary engineering and in continuity with discussions in the metaphysics of science. We first reconstruct which productive activities are involved in contemporary artefact production. From this reconstruction, we derive two general classification systems for artefacts – which we call the 'instrument' and 'product' systems. Then, we adopt from discussions in the metaphysics of science three conditions for classifications to correspond to natural kinds. For each of these three conditions, we discuss which conception or aspect of mind-independence it embodies and to what extent our two classification systems meet it. We conclude that the instrument system is mind-dependent in all ways and the product system only in some. Finally, we identify two options for finding natural classifications of artefacts and develop the second as one that establishes continuity between the metaphysics of science and engineering. This second option leads to a classification system that can correspond to natural kinds and that incorporates the product classes of technical artefacts extensionally.

Keywords Action-theoretical analysis of engineering • Artefact classification • Artefact kinds • Artefacts and mind-dependence • Making and metaphysics

---

W. Houkes (✉)
Philosophy & Ethics, Eindhoven University of Technology,
IPO 1.06, Den Dolech 2, 5600 MB Eindhoven, The Netherlands
e-mail: W.N.Houkes@tue.nl

P.E. Vermaas
Section of Philosophy, Delft University of Technology,
Jaffalaan 5, 2628 BX Delft, The Netherlands
e-mail: p.e.vermaas@tudelft.nl

M. Franssen et al. (eds.), *Artefact Kinds: Ontology and the Human-Made World*, Synthese Library 365, DOI 10.1007/978-3-319-00801-1_10,

10.1 Introduction

Our everyday world is filled with a large variety of human-made items.[1] Almost 150 years ago, hammers came in over 150 types in the Birmingham area alone.[2] In the 1990s, over 110,000 new types of packaged food and beverages were introduced in US supermarkets, which on average contain over 39,000 types of products.[3] Moreover, the classification of these human-made items follows a large variety of guiding principles; hammers, for instance, can be classified in terms of primary purpose, intended use, shape, and many other characteristics.

The general attitude in analytic metaphysics with regard to this multiplicity of classes and types of classes is one of indifference. Classes of human-made items, or of technical artefacts as we will call them, are typically taken as conventional ones that do not correspond to real or natural kinds. Starting point of recent debates on this topic is the observation that technical artefacts result from intentional production and that they are essentially characterised in terms that refer to human purposes and activities. This is taken to create a problem regarding metaphysical status: since artefacts are the products of human activities and are therefore 'mind-dependent', they cannot come in metaphysically respectable kinds; or, even more strictly, they might just be aggregates of real entities (Van Inwagen 1990).

Amie Thomasson (2003, 2007a, b, 2009) has offered a forceful defence of the metaphysical admissibility of so-called human kinds, including artefact kinds. She argues that although human kinds and their instances may depend on human interests and activities, this mind-dependence does not entail conventionalism. Indeed, she claims that this inference reveals a physicalist bias in metaphysics. In the natural sciences, classifications such as the periodic system of chemical elements may correspond to human-independent kinds. Yet in order to inquire, without begging the question, whether human-made items come in kinds, one needs a criterion without a physicalist or mind-independence bias. Thomasson proposes the following:

> for any essentially artifactual kind *K*, something is a *K* only if it is the product of a largely successful intention to make something of kind *K*, where such intentions must involve a substantive conception of what sorts of features are *K*-relevant and this conception largely matches that of some group of prior makers of *K*s (if there are any). (2009, p. 206)

As plausible or open-minded as this criterion may seem, it must be defended against two objections or clarified in two respects.

The first is that intentions may indeed be relevant to the production of artefacts but that it is, on closer analysis of engineering practice, unclear whose intentions in which productive activities are supposed to determine kind membership. Contemporary engineering differs considerably from the artisan model of production implicitly assumed by authors such as Thomasson, Lynne Rudder Baker (2004, 2007)

[1] Throughout the chapter, 'item' refers to objects and processes, which may or may not be human-made; 'class' or 'type' refers to a grouping of items, which may or may not constitute a kind.

[2] Basalla (1988).

[3] Nestle (2002) and FMI (2011).

and Risto Hilpinen (1992, 1993). They take production as the activity of an individual 'artisan' maker or designer, allowing the position that specific intentions of that individual maker or designer determine the kind membership of the artefact he or she creates. Technical artefacts are nowadays typically the result of collective efforts, divided into a number of distinct designing and production phases.[4] A principled reason is needed to seek kind-determining intentions in one or more elements of this elaborate division of labour.[5] Another way to raise this point is to note that items described in the sciences and items described in engineering are treated asymmetrically in analytic-metaphysical discussions. With respect to the former, metaphysicians focus on the question which among the many classification systems correspond to natural kinds; with respect to the latter, they are more concerned with defending that artefacts exist at all, without too much attention for classificatory or engineering practices. Thomasson's criterion might succeed in giving this defence but then raises the question which (if any) of the many classification systems, of say, hammers or packaged foodstuffs, the criterion favours: classifications in terms of primary purpose, in terms of intended use, etc.; or, conversely, which (if not every) of the classification systems it leaves vulnerable to criticisms regarding the respectability of artefact kinds.

A second objection is that, by proposing a criterion specific to artefact kinds, one might trade a question-begging case against such kinds for a question-begging case in favour of them. Advocates of the metaphysical respectability of any contentious classification system could, in principle, propose a criterion that saves realism regarding the classification. To avoid this second type of question-begging, a measure of continuity or resemblance should be established between the criterion for artefact kinds and conditions that have been offered for natural kinds in the sciences. This also allows investigating how the 'mind-dependence' of artefacts and their classification affects the applicability of each condition – in order to make either the best possible case for the metaphysical respectability of artefact kinds or the most damaging case against them.

In this chapter, we present the results of a systematic search for natural kinds of artefacts. In an attempt to address the objections or worries voiced above, it proceeds via an explicit construction of classification systems and an assessment of these classifications in the light of conditions for natural kinds and the role of mind-dependence. We start, in Sect. 10.2, by reconstructing which productive activities are involved in contemporary artefact production. Developing our earlier action-theoretical analysis of using and designing in terms of use plans, we distinguish four productive activities: *designing*, *product designing*, *manufacturing designing* and *making*. This reconstruction adds *make plans* as a central concept to our earlier analysis. Then, in Sect. 10.3, we derive two general classification systems – which we call the 'instrument' and 'product' systems – each with their respective constitutive intentions. In Sect. 10.4, we introduce from the literature three conditions for a

[4] Brown (2000) describes how divisions of labour in productive activities emerged in nineteenth-century engineering in the USA and UK.

[5] This point is elaborated in more detail in Houkes and Vermaas (2009).

natural classification (Bird and Tobin 2008), where a natural classification groups together items in accordance with natural divisions rather than human interests, i.e. groups together items in natural kinds. For each of the three conditions, we discuss which conception or aspect of mind-independence it embodies and to what extent our two classification systems meet it. We conclude that the instrument system is mind-dependent in all and the product system in some ways. Finally, in Sect. 10.5, we present our best case for naturalism with respect to artefact kinds.[6] We identify two options for finding natural classifications of artefacts, corresponding to part of the product classification system. The first option clarifies both Thomasson's view and the question-begging objection to it. The second option puts more stock in establishing continuity between the metaphysics of science and that of engineering. It involves constructing a classification system that can correspond to natural kinds and that incorporates the product system extensionally. We conclude by identifying a remaining challenge.

10.2 A Rational Reconstruction of Engineering Production

In contemporary engineering, production of artefacts (including artificial processes) consists of multistage activities, in a division of labour between and within teams of agents. A presentation in an engineering-design textbook might represent production as starting with a client expressing a goal and, after steps such as goal reformulation, problem decomposition and testing, ending with artefacts to be used by agents. Typically, flowcharts illustrate the reconstructions, accompanied by disclaimers that real engineering-design processes are not as neat as suggested by the box-and-arrow diagrams.[7]

In our own analysis of artefact production, we take as a guiding principle that production is a goal-directed series of activities, where both the series and the different activities involved are subject to standards of (practical) rationality. This makes our result a *rational reconstruction* of production. It may be used as a basis for evaluating the various activities involved and does not fully reflect the complexities of actual practice.

Our reconstruction features four productive activities. In the order of their appearance in the analysis, these are *designing*, *product designing*, *manufacturing designing* and *making*. The activities of designing and product designing can be related to using through the notion of a use plan, as we showed in earlier work.[8] We briefly summarise this analysis immediately below. To extend our earlier results

[6] Following Bird and Tobin (2008), we distinguish kind naturalism (the view that some classifications are natural) from kind realism (the view that natural kinds exist independently of their members or instances).

[7] For example, Hubka and Eder (1988), Roozenburg and Eekels (1995) and Pahl et al. (2007).

[8] Houkes and Vermaas (2010).

to manufacturing designing and making, we have introduced the notion of a *make plan*[9] and have argued that manufacturing designing relates to making as designing relates to using. Designing facilitates using: designing provides use plans by which users can realise their goals; manufacturing designing facilitates making: manufacturing designing provides make plans by which makers can create technical artefacts.

Of the four activities we distinguish, three – product designing, manufacturing designing and making – focus on the production of the item. A central difference with the artisan model of production, as assumed in the work by Thomasson, Baker and Hilpinen, is that these three activities need not be conducted by a single agent. The assumption that the production of artefacts is in the hand of one artisan agent may already be an idealisation regarding pre-industrial times, where divisions of productive labour were not uncommon. In contemporary engineering, teams of agents are standardly involved in product designing, manufacturing designing and making.[10] Members of these teams may change during these activities, and their individual intentions may be incorrect, lost on others or unrelated to the products they are absent-mindedly assembling (Kornblith 2007, sec. 2).

10.2.1 Use Plans

Our inroad to the engineered world is to characterise its contents instrumentally, as items manipulated by agents to realise goals. To analyse the structure and context of these manipulations, we focus on ways of realising goals and introduce *use plans* to spell out 'ways' as ordered series of considered actions; here, use plans can be distinguished from plans in general by their inclusion of interactions[11] with one or

[9] Houkes and Vermaas (2009). There, we called manufacturing designing simply 'manufacturing'; we changed our terminology in this chapter since, in engineering, manufacturing typically refers to both the planning of making and the making itself.

[10] The explicit distinction of separate activities in the production process and the coordination of the resulting division of labour in organisations may be a relatively recent phenomenon (Brown 2000). Engineering definitions of designing typically single out what we have called product designing, yet broader definitions that include use-plan designing can be found (e.g. Hubka and Eder 1988; Roozenburg and Eekels 1995; Brown and Blessing 2005). Manufacturing designing and making are in engineering clearly separated from (product) designing, although (product) designing may include the making of prototypes (Cross 2006, pp. 15–16). This separation is, for instance, described by Visser (2006, p. 160): '[t]he implementation (realization, manufacturing, fabrication, construction) of an artifact based on the specifications that result from the design, is [...] a different task from design, assigned to different professionals'. And Chang (1990, p. 39) takes (product) designing as a process that ends with 'a concept [for a physical entity that functionally satisfies a design requirement] expressed in a communicable media' and manufacturing as a means to realise that concept. Moreover, practices like concurrent engineering show that our conceptual distinction is reflected in an actual division of labour: it is generally acknowledged in engineering that this division of labour has led to new problems, requiring communication and adjustments of the activities by all agents involved.

[11] Interactions with items include physical manipulations (e.g. pressing, turning), remote causal interactions (e.g. accelerating space probes by means of Jupiter's gravitational field) and observations (e.g. watching traffic lights).

more items other than our own bodies. *Using* an item can then be defined as carrying out a use plan for that item. *Designing* is the activity aimed at developing new use plans. Sometimes, a newly designed use plan only involves interactions with existing and readily available items; then, designing is similar to advising or consultancy and mainly involves communicating the new use plan to whomever it may concern. We assume that in designing, the items involved in a use plan are described by the capacities that contribute to the successful execution of the use plan. This description may be enough to identify a suitable item, in case this already exists. In engineering design, the new use plans typically involve interactions with items that do not yet exist or are not readily available. Then, developing and communicating a new use plan is rational only if designing also contributes to making available the items needed. The nonexisting or non-available items involved must be described in more detail than by their contributing capacities alone. Providing this description is the aim of a conceptually separate activity, which we call *product designing*. The specification of items to be produced, in words and images (e.g. blueprints), is typically regarded as the centrepiece of engineering design. Yet we take product designing as an activity that, if it is required at all, is subsidiary to the construction and communication of use plans.

To bring out some details of this reconstruction, let a use plan *up* for an item x be a series of goal-directed, considered actions in which interactions with x are included as contributions to realising the goal g_u of the plan. Let designing be the development of such use plans. Say, a market has developed for effortless nut cracking, giving rise to a goal g_u and a design task of coming up with a use plan *up* to realise the goal.[12] Ordinary nutcrackers require too much effort and may be unsafe, and industrial nut peelers are too noisy and expensive. So, a use plan is developed that makes reference to an item x with a specific capacity ϕ, defining a product-design task of describing x in more basic physical terms than the capacity ϕ. Let *id* refer to this 'item description'. We give general action-theoretical characterisations of designing and of product designing in the Appendix to this chapter, Tables 10.3 and 10.4.

10.2.2 Make Plans

An item does not come into existence by describing it. After product designing, the item should be made. In this part of engineering production, two activities may be distinguished. The first activity, *manufacturing designing*, is aimed at developing a plan for making the described item, taking into account whether the item will

[12] One might want to restrict use plans for an item x to those plans that feature prominent or salient interactions with x. This would prevent standard ways of cooking vegetables from being characterised as use plans for fresh water and ways of cracking nuts as use plans for the nuts as well as the nutcracker. There is no hard and fast rule for determining in general which interactions are sufficiently 'salient': a variety of considerations, such as institutionalisation and regularity of use, come into play when one answers, for an item x, the question "What do you use this for?" or fails to provide a definitive answer, as in the case of fresh water.

be mass-produced, handmade, tailor-made for individual users or even made only once. The second activity, *making*, is aimed at actually producing the item. Manufacturing designing is distinguishable from product designing because even a full description of an item need not specify the actions required to make the item: the described items may be, literally, multiple realisable. Developing a goal-directed series of these actions, a *make plan*, characterises manufacturing designing. This conceptual distinction follows engineering practice, where manufacturing designing has become a profession in its own right. Manufacturing designing is designing: the series of productive actions, aimed at creating an item, involves the manipulation of materials and typically also tools. The make plan for an item x is then a use plan for a set of materials $\{y_1, y_2, \ldots\}$ and possibly a set of tools $\{z_1, z_2, \ldots\}$. Manufacturing designing may give rise to additional product designing, if the developed make plan involves nonexistent materials, tools, machinery or even production plants. Making has in contemporary production become less a matter of specialised training and know-how. In our reconstruction, it amounts to carrying out plans developed in another activity, just like using.

Let *mp* refer to a make plan for an item x with description *id*. Let manufacturing designing be the development of such a make plan on the basis of the item description *id* of x as given by an agent a – typically the product designer. Manufacturing designing is aimed at contributing to the goals of other agents – the makers of x – to create x. So, on the basis of the item description of the household nutcracker, given by a product designer, the manufacturing designer develops a make plan for the nutcracker, which includes manipulations of materials and machinery. The make plan is communicated to makers, who carry out the plan and thus construct and assemble the nutcrackers. Stepwise characterisations of manufacturing designing and making are given in the Appendix, Tables 10.5 and 10.6, again in general terms.

This part of our analysis of contemporary engineering practices brings out another difference with the model of production implicit in analytic metaphysics. Whereas Hilpinen (1992, §VI, 1993, §VI) and Thomasson (2007a, sec. 5) allow intentional selection as a limiting case of making, we explicitly require that making amounts to physically changing materials, either by physical modification of existing materials or by (non-trivial) assembly of those materials.

10.3 Two Classification Systems for Artefacts

There are many ways of grouping together human-made items, using, for instance, terms which refer to intended users ('biker helmet'), primary purpose ('swimsuit'), shape ('U-bend'), material ('wooden shoe'), mode of operation ('nuclear plant') or country of origin ('French press'). Some of these colloquial groupings form more encompassing and systematic classifications than others. More contentiously, some classifications may group together items in accordance with natural divisions rather than (merely) human interests; others may be purely conventional. In the next sections, we will consider how to distinguish natural and conventional classifications of artefacts. In this section, we set the stage by organising and, to some extent, simplifying the subject matter. We show that, on the basis of the distinctions made in the previous section, two comparatively rigorous systems of classification may be constructed: the 'instrument' system and the 'product' system. Both classifications refer to intentional activities and systematise more colloquial ways to group items; yet they differ markedly in their scope and application.

10.3.1 The Instrument System

As reconstructed in Sect. 10.2.1, the activity of designing consists in constructing and communicating a use plan *up* by which agents may realise a goal g_u. In executing this plan, these agents are meant to interact with various items because of the specific contributing capacities ϕ of these items.

This reconstruction of designing allows the construction of a classification system of items. First, the system makes a broad and general distinction between *instruments* and *useless items*. The class of instruments contains every item that is intended to be interacted with in at least one use plan; 'useless' are those items for which there is no use plan. Within the class of instruments, items may be grouped together in two different ways. A first subclassification refers only to the goals g_u to which interaction with the item is meant to contribute. Within these classes of 'goal instruments', a further subdivision can be made by referring to the use plan *up* that involves interaction with the item. An alternative subclassification refers to the goal-contributing capacities ϕ of the items classified as instruments. It can be shown that these contributing capacities ϕ are technical functions.[13] Consequently, a subdivision of the class of instruments according to contributing capacities ϕ yields a function-based subclassification.

This leads to a general classification system (see Table 10.1) where 'goal instruments' and 'plan instruments' make up one subclassification and 'functional instruments' another. We call the entire system the 'instrument system'.

This system applies to almost everything in the 'human-made world': ordinary household utensils; bicycles, trains and aeroplanes; heavy machinery, buildings and industrial installations; and the road system with its bridges and tunnels. The system also extends to the components of these items. The engines, pillars and screws that are parts of machinery, the road system and bicycles, respectively, are items that are interacted with in use plans – not always directly by the end users of these items, but minimally by the product designers who carry out use plans that involve these components when they incorporate the components in their product designs.[14]

Not only items that were created by humans can be called 'instruments'. We sit under trees for shade; wildlife is hunted or observed during safaris; and the stars are used for navigation. Thus, the instrument system applies across the natural-artificial

[13] Vermaas and Houkes (2006) and Houkes and Vermaas (2010).

[14] Vermaas (2006).

Table 10.1 The instrument system

Instruments	Items intended by designers to be interacted with in use plans
g_u goal instruments	Items intended by designers to be interacted with in use plans with goal g_u
up plan instruments	Items intended by designers to be interacted with in use plan *up*
ϕ functional instruments	Items intended by designers to be interacted with in use plans for capacity ϕ

border, where we find such items as wild or domesticated animals and plants, stars and Dutch polders. Moreover, to call something an instrument does not entail that it sees regular or standardised use: the instrument system also applies to pebbles used as improvised paperweights or flat stones that see single-time use in a game of stone skipping.

Subclasses of instruments may be distinguished according to goals – more specifically: according to the goal states of the use plan in which agents are supposed to interact with the items classified. The goal specifications and corresponding classes may vary in grain, from coarse – 'cleaning agents' – to fine – 'candle wax removers'. Even the more fine-grained goal classes of instruments typically contain a wide variety of items, as there are for instance many use plans with the goal of removing candle wax from fabric. One plan involves covering the spot of wax with a brown paper bag and ironing it; another plan involves covering the spot with ice cubes and chipping off the frozen wax; still another includes soaking the spot in carbon tetrachloride. All items interacted with in the execution of these plans may be grouped into the 'candle wax remover' goal class.[15]

As a counterpart to the 'multiple realisability' of most practical goals, most items are multiply usable: they may be interacted with in the course of executing different use plans, with different goal states. Pins, for instance, may be enrolled in the class of 'temporary materials connectors' but also in the class of 'position on map markers'; and brown paper bags and old fashioned flatirons are also in the goal classes of 'grocery container' and 'doorstopper', respectively.[16] The subsystem of overlapping goal-instrument classes reflects many terms by which we refer to artefacts in everyday language: 'container' and 'wax remover' may, for instance, be taken to refer to different but partly coextensive goal classes. However, goal classes should not be taken to extend exclusively to artefacts. A specific natural sea sponge may, for instance, be called a 'cleaning agent' and 'paint applier' – since some painters manipulate them for cleaning their brushes but also for applying paint to walls.

[15] In line with the considerations in n. 12 above, one might want to restrict goal and plan classes to those items that are most 'saliently' interacted with for the purpose or when executing the plan. This would prevent candle wax from being classified as a candle wax remover and the handlebar of a bicycle as a means of transportation.

[16] The example of flatirons used as doorstoppers may be discarded by some as due to recycled or 'improper' use; flatirons may in this vein be taken as only contained in the goal class of 'wrinkle removers'. The other examples are therefore chosen in a way that only proper use is considered.

The use-plan analysis offers a means of subdividing goal classes into plan subclasses. Brown paper bags and irons are included in the 'by-ironing' subclass of the 'wax-remover' class, ice cubes in the 'by-freezing' subclass, etc. A classification on the level of use plans does not lead to unique class membership for the same reason as a classification on the level of goals does not: items typically are interacted with in more than one use plan and therefore belong to multiple *up* plan classes. Even within a goal class, plan subclasses are typically not disjoint: water is interacted with in many plans for cleaning windows; paint features in many plans for decorating walls (as well as all plans for removing paint). One reason for this overlap is that stringent identity conditions may be applied to use plans, on which plans are different as soon as their goal states, included actions or the order of actions included is different.

Alternatively, instruments may be divided into subclasses according to the capacities by which they (are supposed to) contribute to realising goal states. The resulting functional subclasses group together items that have the same contributing capacities, irrespective of the goals for which and the use plans in which they are interacted with. The resulting subclasses are no more disjoint than the goal and plan subclasses. Some knives can be used for their capacity to spread paste-like materials like butter as well as for their capacity to cut materials like bread, which puts them in the functional class of '(butter) spreader' as well as '(bread) cutter'.

There is no neat, general relation between the two ways of subclassifying instruments. On the one hand, functional subclasses are not necessarily more fine-grained than goal/plan subclasses. For instance, with respect to their most obvious intended capacity, knives may be grouped in the 'cutter' functional subclass – which is crude in comparison to classification of knives in terms of goals and use plans (e.g. preparing food, stabbing or performing surgery). On the other hand, goal and plan subclasses are also not more fine-grained than functional subclasses: both the brown paper bag and the iron are included in the 'removing-by-ironing' use-plan class, despite obvious differences in the capacities by which they are meant to contribute to the plan's goal state.

The most fine-grained system of classifying instruments would refer to goal state, use plan and contributing capacities, all as intended by designers. Even this system would not capture all ways in which artefacts are grouped together in ordinary language. There, reference is made to characteristics that play no role in the use-plan reconstruction of designing, such as the size of items ('microprocessor'), their colour ('yellow submarine'), origin ('French press') and brand ('a Ferrari').

10.3.2 The Product System

The instrument system focuses on designer's intentions and functions – features that are often appealed to in philosophical analyses of artefacts. Instruments and their subclasses may therefore seem the only place to look for natural kinds of artefacts or to argue that there are no such kinds to be found. Still, an alternative classification system may be constructed on the basis of our analysis of productive activities – in

Table 10.2 The product system

Products	Items that are created by carrying out make plans
id made-products	Items that are created by makers by carrying out make plans for items fitting description *id*
mp made-products	Items that are created by makers by carrying out a make plan *mp* for items fitting description *id*

particular on the basis of the intentions of manufacturing designers. We call this the *product system.*

As reconstructed in Sect. 10.2.2, manufacturing designers develop plans for making items that fit an 'item description' *id.* This description may be couched in physical, chemical and/or geometrical terms and may refer to both categorical properties and dispositions. It may, moreover, be highly detailed or relatively sketchy. The class of products comprises everything made intentionally, where what matters are primarily the 'creative' intentions of the manufacturing designer, not those of the maker. More specifically, products are items that result from the execution of make plans (by makers) and that are thus intended (by manufacturing designers) to fit some *id.* Similar to Thomasson, we assume that executing a make plan includes a check whether the produced item fits the intended *id* (see the Appendix, Table 10.6). Products are therefore items that are accepted to fit some item description *id.*

The class of products can be subdivided by referring to the *id*'s, leading to subclasses of items that are intended to fit the same *id.* For components and food products, terms that indicate such '*id*' subclasses are common: think of 'PVC pipe' and 'chocolate mousse'. For many other products, *id*'s are available in the form of blueprints or drawings and descriptions in patents. Still, most *id*'s relevant to engineering practice would be too complex to be used in everyday language; proper names such as 'Citroën DS' and 'Aspirin' act as shorthand for such more complex *id*'s.

An even more fine-grained classification is obtained by adding a third, 'made-product' tier to the classification system and specifying productive intentions more fully in terms of make plans *mp.* Since plans are partly identified by their goal state, and having items that fit an item description *id* is the goal state of a make plan *mp*, the resulting classes of *mp* made products are subclasses of the classes of *id* made products. A simple example is the distinction between 'handmade' and 'machine-made' items that may be superimposed on any *id* made-product class. The resulting 'handmade *id*' and 'machine-made *id*' subclasses differ in the actions included in the make plans, but not in the goal state of the plans. Examples that are more relevant to engineering practice would be *mp* made-product classes that are differentiated on the basis of various industrial production processes, e.g. for casting metal objects. Furthermore, product (sub-)classifications that refer to materials, such as 'wooden shoe', can often be reconstructed as *mp* subclasses, since differences in materials used are typically reflected in differences in actions included in the make plan. Not all ordinary-language classifications of products may be reconstructed in this way: origin-based classifications ('French press'), for instance, cannot be derived from differentiating features of make plans.

This leads to the product classification system depicted in Table 10.2.

The class of products includes, like that of instruments, ordinary household utensils; bicycles, trains and aeroplanes; heavy machinery, buildings and industrial installations; and the road system with its bridges and tunnels. It also extends to many components, such as engines, pillars and screws, which are intentionally created in accordance with relatively autonomous make plans: they may be intended to be used as components of many different items; and execution of the make plans for components need not be spatially or temporally contiguous with the execution of the make plan of the items of which they are components. Moreover, similarity to a naturally occurring item is no reason to exclude something from the class of products. Water that is intentionally purified or intentionally made by burning hydrogen may be called a product; the same goes for artificial vitamins and, arguably, many domesticated animals and plants.[17]

The product and instrument classes do not coincide. Planets and pebbles, for instance, cannot be called 'products': our practical interactions with these items may not be reconstructed as execution of a make plan, but only of a use plan. Conversely, the class of products largely but not entirely lies within that of instruments, since most – but not all – make plans are constructed and executed in order to create items that are to be manipulated in carrying out use plans. Occasionally, however, manufacturing and making may be unrelated to use plans. Scientific or purely creative challenges may be candidates for such non-instrumental production activities. For instance, creating transuranic elements, artificial molecules such as Buckminsterfullerene, or synthetic organisms, just for showing that it can be done, may require the construction and carrying out of elaborate make plans. So, typically, do more quotidian challenges such as building the largest sandcastle on the beach. The goal state of these make plans is to have items that fit highly specific item descriptions *id* but that are not intended to be interacted with in any familiar use plan.

10.4 From Classifications to Mind-Independent Kinds

In the previous section, we presented two classification systems for artefacts, both based on reconstructions of contemporary engineering practices. We now turn to the question which of these systems, if any, may be taken as a natural classification.

We approach this question in the 'metaphysics of engineering' in continuity with the metaphysics of science. There, a central issue is to identify which, if any, scientific classification systems are natural, i.e. independent from human interests and conventions. Mendeleev's periodic system of the elements is a prime candidate for such a natural classification, whereas the classification of animals in the *Celestial Emporium of Benevolent Knowledge*, or the classification of baseball players by their fielding

[17] The make-plan analysis determines the focus of the arguments: a domesticated plant or animal may be called a 'product' if breeding it may be reconstructed as making – so, if there is a constructed and communicated make plan for the plant or animal. Genetic engineering of organisms can be reconstructed as such, and so can carrying out a deliberate breeding scheme; merely selecting plants or animals for desired properties cannot.

positions, is not. Since independence from human interests can, obviously, not be established directly, several more proximate conditions have been proposed for determining which classification systems might be regarded as natural. There appears to be no consensus about the list of relevant conditions, some of which are mostly used implicitly rather than stated explicitly. Moreover, the situation is complicated because several conditions have been brought forward in order to attack the very idea of natural classifications or natural kinds (e.g. by De Sousa 1984). Here, we focus on three conditions that are listed in an overview of work on natural kinds (Bird and Tobin 2008)[18]:

1. *Intrinsicality*: Members of a natural kind share some or all of their intrinsic properties.
2. *Distinctness*. Natural kinds must be categorically distinct from each other.
3. *Stability*. Items cannot, after their creation, gain or lose membership of natural kinds without changes in intrinsic properties.

In this section, we apply these conditions to the instrument and product classification systems. Our primary purpose in doing so is to examine whether one of these systems may be regarded as 'natural' on the basis of these conditions. If either system is, a non-question-begging case is made for the metaphysical respectability of artefact kinds. No specific criterion for the latter would be needed, since there would exist shared standards for the metaphysics of science and that of engineering. The standard might still be controversial, but this controversy would equally affect claims regarding natural classifications in science and in engineering.

The various conditions on natural classifications also appear useful with regard to specifying mind-dependence because they, in various ways, make explicit the idea that natural classifications are independent of 'social and psychological facts about human beings' (Hacking 1991) or 'objective' (De Sousa 1984). Failure to meet a condition therefore may correspond to a specific kind of mind-dependence.

Below, we review, for each of the three conditions, whether it is satisfied by (parts of) the instrument and product classification systems and to what extent a failure to do so indicates the mind-dependence of artefacts and their classification.

10.4.1 The Intrinsicality Condition

The Intrinsicality condition is based on the intuition that an item's membership of some classification systems is determined by properties that the item has irrespective of its relation to, or even the existence of, any other items. The atomic number

[18] Other conditions for natural classifications offered in the literature, but not considered here, include the following: that natural kinds support inductions or even laws of nature (Bird and Tobin 2008), that natural kinds form a kind (*ibid.*), that kind membership is unique (De Sousa 1984), that use of natural kinds is significant to the development of human knowledge (Hacking 1991) and that natural kinds are associated with real essences (Ellis 2001).

of chemical elements is a paradigmatic example of such an intrinsic property, whereas the human state of being jealous is not. Failure to meet Intrinsicality does not entail that items are mind-dependent: weight, for instance, is dependent on an external gravitation field and therefore not an intrinsic property, but it is not mind-dependent. Still, one might think that *because* artefacts depend on human intentions for their existence and their most characteristic properties, classifications of artefacts are bound to fail an Intrinsicality test.

To avoid jumping to conclusions, we examine in some detail whether the instrument and product classification systems group together items in terms of intrinsic properties. Both systems contain subclasses of which membership is determined by particular properties. Yet for most subclasses, these are properties of human activities, not of the items involved in them. For instance, an item belongs to a goal subclass in case someone intends it to be used for a particular purpose. Likewise, both use-plan and make-plan subclasses may group together items by characterising goal-directed, ordered series of intentional actions, which include interactions with the items. Returning to earlier examples, an *up* plan-instrument subclass may group together everything used to remove candle wax by ironing, and items in an *mp* made-product subclass may all have been made by hand. Thus, these classifications are strongly dependent on human activities in the sense that they group together items in terms of such activities.

Still, both classification systems include subclasses that, at least at first glance, characterise the properties of items, and not those of activities – namely, functional instruments that share a capacity ϕ and *id* made-products that share an item description *id*. On closer inspection, the capacity ϕ fails to specify an intrinsic property. Not only may the item fail to have or manifest ϕ, but even if it has or manifests ϕ, this depends on human beings interacting with it (or just intending to interact with it) in specific ways. An item is not a member of the functional instrument subclass of, say, water tanks in virtue of, among other things, its impermeability, but because it is intended to be manipulated in specific ways for the purpose of containing water. Not all water tanks are in fact impermeable, and even for those that are, or would be if used, their impermeability does not in itself determine class membership, although it may explain why their possible use for the purpose would be successful – the kerosene tank of Air Force Number One is impermeable, but not a member of the functional instrument subclass of water tanks, since it is not to be expected that someone has developed the associated use plan for the tank.

By contrast, *id* made-products appear to share intrinsic properties. By definition, membership of these classes is determined by the item description *id*, which consists of categorical or dispositional, physical, chemical or geometrical properties – such as size, material composition, conductivity and/or permeability. In principle, items that are intentionally produced can be identified independently of any considered use. Thus, 'marble sphere with ten-centimetre radius' identifies a class of *id* made-products irrespective of any interaction with, including purposeful manipulation by, human beings.

Despite first appearances, however, membership of *id* made-product classes is not mind-independent. For one thing, all items in these classes are created intentionally. A ten-centimetre radius marble sphere formed by some freak geological event would not qualify as a member of the *mp* made-product class. Moreover, not nature but the decisions of product and/or manufacturing designers determine which properties feature in an *id*. The items in the subclass may share other properties: all considered marble spheres with a radius of ten centimetres may, for instance, be made with marble from the same quarry – but only accidentally so unless this is specified in the defining *id*, the contents of which are determined in the human activities of product designing and manufacturing designing.

Summing up, both the instrument and the product classification systems are revealed as mind-dependent with respect to the Intrinsicality condition, albeit in different ways.

10.4.2 The Distinctness Condition

The Distinctness condition features on several lists offered in the literature, e.g. in De Sousa's (1984) condition of 'sharp boundaries' and Hacking's (1991) 'definability'. It is motivated by the intuition that membership of natural kinds cannot be indeterminate, since that would require human beings to arbitrate to which of two gradually merging kinds an item belongs. Thus, failure to meet Distinctness entails a specific kind of mind-dependence of classes – conventional or subjective boundaries – which would undermine the naturalness of a classification system.

Consider, in this light, goal classes and functional classes in the instrument system. The 'removing wax' and 'removing grease' goal classes or the 'storing milk' and 'heating milk' goal classes may be readily distinguished: few items are, intuitively, in grey areas between these classes, so that arbitration seems uncalled for. However, other goal classes do merge gradually into each other. Some items, for instance, are as much playthings as educational materials (or are at least advertised as contributing to both entertainment and education), whereas other items are quite clearly only for play or for education. The reason is, of course, that some human activities serve multiple purposes or achieve one thing by striving for another. In many such cases, neither goal is subsidiary to the other and neither requires realisation of the other. If a nonarbitrary line between the resulting, gradually merging goal classes can be drawn at all, it would require a close examination of individual human motivations.

For functional classes, the indistinctness of the capacities ϕ presents additional problems. As alluded to above, these capacities may be described in both coarse-grained and fine-grained terms. Most of us would, for instance, describe the capacity for which a tablet of Aspirin is most regularly used as 'alleviating pain'. Pharmacologists might (or should), however, replace such a coarse-grained, goal-directed description with one that specifies how the tablet and its ingredients are intended to affect physiological mechanisms. One might even specify different classes of painkillers, in terms of more fine-grained, physiologically distinct capacities. Such distinctions, or the need for such distinctions, depend not only on the availability of relevant knowledge (which affects every classification) but also on the social need for such knowledge. Drawing the line between various kinds of painkillers is unnecessary for most users, and its necessity for experts arises from the intended usage (e.g. facilitating regular use without side effects or predicting overdose effects) rather than intrinsic properties of the items. Alcoholic beverages may, for instance, be distinguished according to precise physiological effects just as painkillers – but their usage does not typically require such distinctions.

Turning to the product system, *id* made-product classes appear to be sufficiently distinct. The item descriptions *id* may contain vague terms, such as 'brittle', 'long' or 'sharp'. However, as the reconstruction of manufacturing designing makes clear, this activity requires an agent to specify vague terms in the description *id* at least up to the point that the results of making can be evaluated. Hence, *id*'s should be definite enough to facilitate checks in manufacturing designing: classes of '*id* made products' therefore single out kinds that are just as distinct. On closer inspection, the stringency with which these distinctions are imposed reveals a further dependence on human interests. Items used in official sports tournaments are, for instance, subject to highly specific *id*'s, whereas far less specific *id*'s apply for similar items used in playgrounds. Furthermore, the *id*'s of many made products contain continuously variable properties, such as length or amount of an active ingredient. Whether and how these are made discrete in *id*'s is strongly interest dependent. PVC pipes, for instance, come in many lengths, which would seem relevant in the *id*'s that guide their production. This would mean that PVC pipes come either in as many distinct kinds as they are applied for or in one kind that is indistinct in one aspect of its *id*. Tablets of Aspirin, to give another example, come in several distinct dosages of acetylsalicylic acid. However, the distinctness of this aspect of their *id* depends on distinct applications by human beings (e.g. for children, for adults, for daily use, for incidental use), not on any distinction 'in nature'. Furthermore, professionals and experts may specify properties in the *id* such as 'toughness' or 'flexibility' in less coarse-grained terms, because there is a practical interest in doing so.

Summing up, both the instrument and the product system are lacking in terms of Distinctness. These lacks reflect various dependencies of artefact classification on human interests. Some items are classified in indistinct instrumental classes because we value their usefulness for multiple purposes; and the distinctness of any artefact classification depends on our interest in specifying it. For instruments, such a specification would involve determining the relevant intentions or actions, revealing a rather fundamental mind-dependence. For *mp* made-product classes, there is also a fundamental mind-dependence in the (specification of the) productive actions that determine class membership.

10.4.3 The Stability Condition

The Stability condition for natural classifications is controversial. It is explicitly required by some (e.g. De Sousa, who goes on to show that no classification system is in fact stable), implicitly appealed to by others (e.g. Bird and Tobin) and mentioned as an optional feature by still others (e.g. Ellis, who distinguishes between fixed and variable natural kinds). Like Intrinsicality, failure to meet Stability does not entail that a classification system depends on human interests but just that items may lose membership of some classes or gain that of others. However, for artefacts, one may suspect that this dynamics is not intrinsic but reflects changes in human interests.

The class of instruments and its subclasses confirms this suspicion, since they depend on the demands of users and on the state of the art in engineering design. In modern engineering practice, this state of the art can change quickly. In particular, the constant construction and adoption of new use plans leads to an intricate dynamics of instrumental subclasses. Membership of the different instrumental subclasses is highly context dependent, where common knowledge and acceptance of use plans provide much of the relevant context. Items that are members of one subclass (goal, plan or functional) may become members of other subclasses, even without physical modification. Tablets of Aspirin have, for instance, long been members of the goal and functional class 'painkiller', but have recently also become members of the goal and functional class 'blood-thinner' and may be on the verge of being enrolled in that of 'anticarcinogenic'. Conversely, technological developments may make use plans obsolete, causing items to lose membership of instrumental subclasses or, in rare cases, of the class of instruments altogether. A case of subclass dropout is plate armour, which went from the goal and functional class 'protective gear' in mediaeval times to 'ceremonial garb' in the seventeenth and eighteenth centuries and 'decoration' or 'museum exhibit' in the present age.

The product system is more stable. In particular, *id* made-product classes single out items by the item description *id* that is the goal of the make plan that led to the item. Therefore, an item retains membership of an *id* made-product class throughout its lifetime. Embedding it in a different context of use, i.e. including interactions with the item in a use plan that did not prompt the construction of its make plan, is irrelevant to this: it changes or adds to the item's functionality, without changing the item's membership of its original *id* made-product class. Tablets of Aspirin are, for instance, created as members of 'tablets-containing-*x*-milligrams-of-acetylsalicylic-acid' subclasses. This remains the case after Aspirin is recognised to reduce blood clotting or to prevent tumour growth and is included in new goal-instrumental subclasses. Tablets of Aspirin would even remain in their original *id* made-product classes if any or all of their uses would become obsolete and they would consequently no longer be called 'painkillers' or even useful.

Furthermore, since executing a make plan, on our reconstruction, requires at least one physical modification of the item or of its constitutive materials, Cambridge changes are ruled out: something cannot lose membership of an *id* made-product class without a modification of at least one of its intrinsic properties. However, not every modification counts. Ordinary wear and tear will affect some of the properties of an artefact, and some of these changes may be compatible with the artefact remaining to meet its original *id*, meaning that it does not lose its *id* made-product class membership: friction changes the profile of tyres and the sharpness of knives, but these changes may be within set limits. Yet eventually the changes may be such that an artefact would not meet its original *id* and that may even be after a short period of use. The item then loses membership of its original *id* made-product class. In this case, it may be still re-made, meaning that there will be a new cycle of manufacturing designing and making, and possibly also of plan and product designing – for instance, in maintenance, refurbishment and recycling. Here, somewhat paradoxically, the stability of class membership despite changes in properties reveals another aspect of the mind-dependence of the product classification system – namely, human expectations regarding changes in or even loss of the properties that determine class membership.

10.5 Two Ways to Find Natural Kinds of Artefacts

The discussion in the previous section shows that neither the instrument nor the product classification system of artefacts has a strong claim to being natural. The instrument classification system does not meet any of the three conditions considered. The product classification meets Stability at the level of *id* made-product classes, and it comes close to meeting Distinctness and Intrinsicality at the same level – but coming close does not count in analytic metaphysics. As discussed, these failures reflect in various ways how artefacts and their classifications depend on human interests and activities.

Still, we argue in this section that there are at least two options for maintaining the view that, despite their mind-dependence, artefacts come in natural kinds.

A first option is to acknowledge that the mind-dependence of artefacts makes it impossible for their classifications to meet all three conditions but that some classifications of artefacts may still be taken as natural. More specifically, one might argue that one or more of the conditions considered in the previous section should be relaxed to accommodate one or more systems of artefact classification. Thomasson's view, presented in Sect. 10.1, embodies this option: it involves weakening the Intrinsicality condition by allowing that intrinsic properties may be related to productive intentions. Thomasson defends this by accusing the original Intrinsicality condition of a physicalist bias and therefore of begging the question against the metaphysical acceptability of artefact kinds. Rephrased in terms of our own results, Thomasson proposes to take as a natural classification the grouping of artefacts into *id* made-product classes. This classification meets the modified Intrinsicality condition, and our make-plan analysis adds a more precise identification of the constituting intentions of kinds as the intentions that come with make plans.

This option and its implementation need not or should not convince orthodox analytic metaphysicians, for two reasons. First, the original Intrinsicality condition spells out the basic intuition behind distinguishing natural and conventional classifications, an intuition that is lost by weakening this condition. This makes the new Intrinsicality condition vulnerable to the charge that it, and not the original condition, begs the question and that it does so in favour of natural classifications of artefacts – specifying the charge against Thomasson's view presented in Sect. 10.1. Second, whatever changes are made to the Intrinsicality condition, the *id* made-product classification still only approximately satisfies the Distinctness condition. Hence, those taking the first option should weaken, adjust or challenge two conditions, both of which make explicit in different ways the idea that natural classifications should be mind independent.

We therefore present a second option for finding natural classifications of artefacts, which does not involve weakening the Intrinsicality condition.

Let us return to the analysis given in the previous section about how *id* made-product classes violate the Intrinsicality condition. As noted there, membership of an *id* made-product class is not determined by intrinsic properties but by the make-plan-constitutive intentions to produce items with intrinsic properties – let us call these properties *p* – that are included in the item description *id*. We can now observe that, at least for moderately complex technical artefacts, the *id* made-product class has the same extension as the class of items that have the properties *p* specified by the *id* of the class. For, as Reverend William Paley famously pointed out in another context, there simply are no 'swamp watches' (or 'heath watches'), i.e. there are no naturally occurring items that have the properties *p* that are included in the *id* of any class of watches.

This observation allows the introduction of a classification system of artificial items that share specific intrinsic properties *p*. For each class of *id* made products, there is a corresponding class to which items belong because they have intrinsic properties *p* contained in the item description *id*. By construction, this '*p* classification system' meets the Intrinsicality condition: the intrinsic properties *p* determine class membership. Moreover, the *p* system meets the Stability condition in largely the same way as the *id* made-product classification system: items are members of a *p* kind if and only if they have intrinsic properties *p*, and changes in the intrinsic properties of items cause items to lose or gain membership of specific *p* kinds.

These considerations do not establish that the product classification system of technical artefacts is natural. Rather, we have shown that, for *id* made-product classes of moderately complex technical artefacts, there are classes with the same extension that satisfy Intrinsicality and Stability. This allows no more (or less) than the conclusion that the classes of aeroplanes, watches and other non-naturally occurring items are 'natural by proxy'. *Id* made-product classes of other, less-complex artefacts are extensionally *included* in *p* classes. Members of such classes can come into existence by either natural processes or intentional production. Examples are water, Aspirin and marble spheres (assuming that freak geological phenomena can create such spheres).

The natural sciences may already group together naturally occurring members of such classes in terms of their intrinsic properties *p*. What engineering, reconstructed as in Sect. 10.2, adds are *non-naturally occurring* items with intrinsic properties *p*, where these properties can be the same as the intrinsic properties *p* acknowledged by the sciences for naturally occurring items. Now, if one takes the scientific groupings to correspond to natural kinds, it seems parochial to exclude the non-naturally occurring items as members of these natural kinds: if one takes – as per the Intrinsicality condition – the intrinsic properties *p* as determining kind membership, excluding some items that have *p* just on the basis that they were created to have *p* is a question-begging exception to one's own criterion of admittance. Taking this argument one step further, it is equally parochial to exclude *non-naturally occurring* items with intrinsic properties *p* that are different from the intrinsic properties *p* acknowledged by the sciences for naturally occurring items. Non-naturally occurring items define *p* classes that meet Intrinsicality just as much as *p* classes of naturally occurring items; excluding the former *p* classes to correspond to natural kinds just on the basis of that they contain mainly or only members that were created to have *p* is again a question-begging exception to the criterion of admittance.

This second option for arguing that artefacts come in real kinds also goes some way towards meeting Distinctness. Physical and chemical *p* kinds are, arguably, sufficiently Distinct, whether their members are naturally occurring or not. By contrast, human interests may determine how the boundaries of artificial *p* kinds with more complex properties are identified. It is not immediately clear, however, that a specific problem regarding artefact kinds remains here. First, debates regarding metaphysical indeterminacy (e.g. Van Inwagen 1990) or epistemic and semantic vagueness (e.g. Williamson 1994) may simply take problems regarding the boundaries of artificial *p* kinds in stride. Moreover, one might draw a parallel between artificial *p* kinds and biological species. Arguably, the latter are natural kinds, but their boundaries are not sharp, and drawing boundaries may involve discretisation of continuously varying properties – like what was discussed as a problem for some artefact classes in Sect. 10.4.2 above. Again, the interest dependence of the discretisation of (a limited number of) artificial *p* kinds may not present a specific problem. Thus, problems with regard to Distinctness may not arise for all artificial *p* kinds and, even when they do, may not be specific to such kinds.[19]

Supposing that Distinctness can be met along one or more of these lines, the second option discussed here leads to the following combined picture of kinds in the natural sciences and engineering.[20]

On the one hand, one may distinguish natural kinds in terms of the theories or disciplines that deal with their intrinsic properties *p*: physical kinds (e.g. elementary particles), chemical kinds (e.g. chemical elements and complex molecules), biological kinds (e.g. species) and artificial kinds (e.g. specific aeroplanes and watches). On the other hand, irrespective of controversies regarding the distinctions between these natural kinds or regarding specific examples of natural kinds, one may distinguish natural kinds in terms of the processes that bring its members into existence: by naturally occurring processes alone, by intentional production alone or by either. The third type extensionally includes *id* made-product classes, and the second type

[19] We would like to thank Maarten Franssen and Thomas Reydon for suggesting some of these ways of meeting the Distinctness condition.

[20] Our focus on natural kinds identified in physics, chemistry and biology should not be taken as expression of a view that other disciplines, such as the behavioural, cognitive and social sciences, do not identify natural kinds.

coincides with them; only the first has no overlap with any *id* made-product class. Technological advances make it increasingly difficult to find examples of 'purely naturally occurring' kinds – although creating members of the barred-spiral-galaxy kind lies far beyond presently conceivable human capabilities. Changes in these capabilities also continue to reveal new natural kinds, such as transuranic elements and Buckminsterfullerene molecules; and, less dramatically, it makes it possible to intentionally create members of existing natural kinds, such as diamond and acetyl-salicylic acid.

On this second option, technical artefacts have a relatively straightforward status within analytic metaphysics while staying close to orthodox intuitions about natural kinds: *id* made-product classes, for which intentional production is necessary to realise members of the classes, are extensionally equal to related *p* natural kinds; and *id* made-product classes, for which members can also come into existence by other processes than intentional production, are extensionally within related *p* natural kinds. All technical artefacts that are the result of carrying out a make plan are therefore members of *p* natural kinds and are thus to be taken as respectable within analytic metaphysics. Moreover, if weak realism for natural kinds is accepted as defined by Bird and Tobin (2008), then all these technical artefacts are also members of real kinds.

Acknowledgments Our thanks go to Maarten Franssen, Peter Kroes and Thomas Reydon for their comments on an earlier draft of this chapter. Research by Wybo Houkes and research by Pieter Vermaas were supported by the Netherlands Organisation for Scientific Research (NWO).

Appendix: Action-Theoretical Descriptions

Table 10.3 Designing

D.1	The designer d wants to contribute to a user's goal of bringing about a state g_u
D.2	d believes that the state g_u' is the closest consistent and viable approximation of g_u and intends to contribute to bringing about g_u'
D.3	*d* believes that if the users follow an appropriate use plan *up* that involves the interaction with items $\{x_1, x_2, \ldots\}$, this will contribute to bringing about g_u' and intends to construct this plan *up*
D.4	*d* intends to contribute to producing the items x_i, x_j, etc., that do not yet exist by *product-designing* them
D.5	*d* intends to communicate *up* to the users
D.6	*d* believes that g_u' can or cannot be brought about by the users to whom *up* is communicated. This belief is based on the observation that some users go through a sequence of actions *up'* and bring about g_u'' and on a comparison of g_u'' with g_u'
D.7	*d* decides that her goal to contribute to bringing about g_u' has been achieved or not. In the latter case, *d* may decide to repeat the entire *D* sequence, settle on another plan (return to *D.3*), re-product-design at least one of the items x_i, x_j, etc., (return to *D.4*) or re-attempt communication (return to *D.5*)

Table 10.4 Product designing

PD.1	The designer *d* believes that an item *x* with physicochemical capacity ϕ does not exist
PD.2	*d* intends to contribute to realising the goal state g_{dx}, consisting of the existence of a description *id* of an item *x* with physicochemical capacity ϕ
PD.3	*d* intends to describe an item *x* with physicochemical capacity ϕ for realising g_{dx}
PD.4	*d* believes that a composite of the components $\{c_1, c_2, \ldots\}$, where c_1 has capacity ϕ_1, c_2 has capacity ϕ_2, etc., has the desired capacity ϕ[a]
PD.5	For each component c_i, *d* intends to contribute to bringing about the goal state g_{dci}, consisting of the existence of a description of an item c_i with capacity ϕ_i; if *d* believes that this item c_i already exists, then design task g_{dci} is fulfilled by describing this item. If *d* believes that this item c_i does not exist, then another decomposition step *PD.4*, component design task step *PD.5* and integration step *PD.6* are made in order to fulfil design task g_{dci}
PD.6	*d* believes that the various design tasks g_{dci} are fulfilled simultaneously, i.e. that the item *x* composed of the described components $\{c_1, c_2, \ldots\}$ has the capacity ϕ
PD.7	*d* intends to communicate the description *id* of item *x*, possibly along with instructions for production and assembly of it and/or its components to appropriate agents

[a]In principle one can have only one component, in which case *x* is equal to c_1 and ϕ is equal to ϕ_1

Table 10.5 Manufacturing designing

MD.1	The manufacturing designer *md* wants to contribute to a maker's goal of bringing about an item *x* as described with description *id* by the agent *a*
MD.2	*md* believes that an item *x′* is the closest consistent and viable approximation of *x* and intends to contribute to bringing about *x′*
MD.3	*md* believes that if the makers follow an appropriate make plan *mp* that involves the manipulation of materials $\{y_1, y_2, \ldots\}$ and tools $\{z_1, z_2, \ldots\}$, this will contribute to bringing about *x′* and intends to construct this plan *mp*
MD.4	*md* intends to contribute to producing the items y_i, y_j, etc., that do not yet exist by *product-designing* them[a]
MD.5	*md* intends to communicate *mp* to the makers
MD.6	*md* believes that *x′* can or cannot be brought about by the makers to whom *mp* is communicated. This belief is based on the observation that some makers go through a sequence of actions *mp′* and bring about *x″* and on a comparison of *x″* with *x′*
MD.7	*md* decides that her goal to contribute to bringing about *x′* has been achieved or not. In the latter case, *md* may decide to repeat the entire *MD* sequence, settle on another plan (return to *MD.3*), re-product-design at least one of the items y_i, y_j, z_k, z_l, etc., (return to *MD.4*) or re-attempt communication (return to *MD.5*)

[a]If, say, the right machines for making *x* singled out by the manufacturing designer *md* do not yet exist, he or she may decide to let them be made first. This turns making into an iterative process: items are made with items that were previously made

Table 10.6 Making

M.1	The maker *m* wants to bring about the existence of an item *x* as described by description *id* by an agent *a*
M.2	*m* either chooses from a set of available alternatives a make plan *mp* for bringing about the existence of *x* that involves the manipulation of materials $\{y_1, y_2, \ldots\}$ and tools $\{z_1, z_2, \ldots\}$
M.3	*m* believes that the physical circumstances support realising *mp* and that he possesses the necessary skills
M.4	*m* intends to carry out *mp* and acts accordingly
M.5	*m* observes x' as the outcome of *mp* and compares x' with *x*
M.6	*m* believes that *x* has been brought about or not. In the latter case, he may decide to repeat the realisation of *mp* or to repeat the entire *M*-sequence

References

Baker, L. R. (2004). The ontology of artefacts. *Philosophical Explorations, 7*, 1–14.

Baker, L. R. (2007). *The metaphysics of everyday life: An essay in practical realism*. Cambridge: Cambridge University Press.

Basalla, G. (1988). *The evolution of technology*. Cambridge: Cambridge University Press.

Bird, A., & Tobin, E. (2008). Natural kinds. In *Stanford encyclopedia of philosophy*. http://plato.stanford.edu/entries/natural-kinds. Accessed May 2012.

Brown, J. K. (2000). Design plans, working drawings, national styles: Engineering practice in Great Britain and the United States, 1775–1945. *Technology and Culture, 41*, 195–238.

Brown, D. C., & Blessing, L. (2005, September 24–28). The relationship between function and affordance. In *Proceedings of the ASME 2005 IDETC/CIE conference*, Long Beach, Paper no: DECT2005-85017.

Chang, T. C. (1990). *Expert process planning for manufacturing*. Reading: Addison-Wesley.

Cross, N. (2006). *Designerly ways of knowing*. London: Springer.

De Sousa, R. (1984). The natural shiftiness of natural kinds. *Canadian Journal of Philosophy, 14*, 561–580.

Ellis, B. (2001). *Scientific essentialism*. Cambridge: Cambridge University Press.

FMI. (2011). *Food marketing institute supermarket facts 2010*. http://www.fmi.org/research-resources/supermarket-facts. Accessed January 2013.

Hacking, I. (1991). A tradition of natural kinds. *Philosophical Studies, 61*, 109–126.

Hilpinen, R. (1992). Artifacts and works of art. *Theoria, 58*, 58–82.

Hilpinen, R. (1993). Authors and artifacts. *Proceedings of the Aristotelian Society, 93*, 155–178.

Houkes, W., & Vermaas, P. E. (2009). Contemporary engineering and the metaphysics of artefacts: Beyond the artisan model. *The Monist, 92*, 403–419.

Houkes, W., & Vermaas, P. E. (2010). *Technical functions: On the use and design of artefacts*. Dordrecht: Springer.

Hubka, V., & Eder, W. E. (1988). *Theory of technical systems*. Berlin: Springer.

Kornblith, H. (2007). How to refer to artifacts. In E. Margolis & S. Laurence (Eds.), *Creations of the mind: Theories of artifacts and their representation* (pp. 138–149). Oxford: Oxford University Press.

Nestle, M. (2002). *Food politics*. Berkeley: University of California Press.

Pahl, G., Beitz, W., Feldhusen, J., & Grote, K. (2007). *Engineering design. A systematic approach* (3rd ed.). London: Springer.

Roozenburg, N. F. M., & Eekels, J. (1995). *Product design: Fundamentals and methods*. Chichester: Wiley.

Thomasson, A. M. (2003). Realism and human kinds. *Philosophy and Phenomenological Research, 67*, 580–609.

Thomasson, A. M. (2007a). Artifacts and human concepts. In E. Margolis & S. Laurence (Eds.), *Creations of the mind: Theories of artifacts and their representation* (pp. 52–73). Oxford: Oxford University Press.

Thomasson, A. M. (2007b). *Ordinary objects*. Oxford: Oxford University Press.

Thomasson, A. M. (2009). Artefacts in metaphysics. In A. W. M. Meijers (Ed.), *Handbook of philosophy of science. Vol. 9: Philosophy of technology and engineering sciences* (pp. 191–212). Amsterdam: Elsevier.

Van Inwagen, P. (1990). *Material beings*. Ithaca: Cornell University Press.

Vermaas, P. E. (2006). The physical connection: Engineering function ascriptions to technical artefacts and their components. *Studies in History and Philosophy of Science, 37*, 62–75.

Vermaas, P. E., & Houkes, W. (2006). Technical functions: A drawbridge between the intentional and structural nature of technical artefacts. *Studies in History and Philosophy of Science, 37*, 5–18.

Visser, W. (2006). *The cognitive artifacts of designing*. Mahwah: Lawrence Erlbaum.

Williamson, T. (1994). *Vagueness*. London: Routledge.

Chapter 11
Artefactual Systems, Missing Components and Replaceability

Nicola Guarino

Abstract A very important practical concern of modern information systems is to make explicit, for the purpose of mutual understanding and interoperability, people's assumptions about everyday reality. This is one of the reasons behind the emergence of applied ontology as an interdisciplinary area of research, which builds on the powerful tools of formal ontology and the insights of philosophical investigation to provide useful, cognitively transparent and computationally manageable formal models, also known as *(computational) ontologies*. The importance of such ontologies is increasingly recognised nowadays, as a complement to standard technical documentation, especially in the (collaborative) design, production, maintenance and deployment of complex technical artefacts. Under this perspective, I will focus on an ontological puzzle that is still lively debated. When engineers or technicians speak of technical things and discuss about construction or maintenance problems, they tend to ascribe a genuine ontological status to their 'creatures', even if they do not have a physical presence. This seems to be a systematic phenomenon in the case of *system components* that are temporarily missing or undergo replacement. A technician would, for instance, talk about a cable that connects to a lamp in a smashed headlamp of a car, thus referring to the lamp even if it is not there anymore. This chapter will offer a practitioner-oriented ontological account of such situations, concerning objects playing the role of functional components in larger artefactual systems. I will argue that the way people refer to such objects presupposes a non-standard ontological behaviour, allowing for complete replacement and virtual presence.

Keywords Artefacts • Artefactual systems • Computational ontologies • Design specification • Direction of fit • Functional roles • Intentional selection • Replaceability • System components

N. Guarino (✉)
ISTC-CNR, Via alla Cascata, 56/C, 38123 Trento, Italy
e-mail: nicolguar@gmail.com

M. Franssen et al. (eds.), *Artefact Kinds: Ontology and the Human-Made World*, Synthese Library 365, DOI 10.1007/978-3-319-00801-1_11,

11.1 Introduction

A very important practical concern of modern information systems is to make explicit, for the purpose of mutual understanding and interoperability, people's assumptions about everyday reality. This is one of the reasons behind the emergence of applied ontology (Guarino and Musen 2005; Munn and Smith 2010) as an interdisciplinary area of research that builds on the powerful tools of formal ontology and the insights of philosophical investigation to provide useful, cognitively transparent and computationally manageable formal models, also known as (*computational*) ontologies (Guarino 1998). The importance of such ontologies is increasingly recognised nowadays, as a complement to standard technical documentation, especially in the (collaborative) design, production, maintenance and deployment of complex technical artefacts. In this context, I would like to focus on an ontological puzzle that is still lively debated (West 2003, 2010). When engineers or technicians speak of technical things and discuss their construction or maintenance problems, they tend to ascribe a genuine ontological status to their 'creatures', even if they do not have a physical presence. This sometimes happens when they speak of new product models that are yet-to-be as real things and definitely seems to be a systematic phenomenon in the case of missing system components. Suppose, for instance, that the right headlamp of my car has been smashed in an accident and has been removed from its socket in order to be replaced. The lamp is not there anymore, but still a technician would say that *this lamp* has been replaced twice already and that it is perhaps useful to also replace the cable that connects to *it*. To what do *the lamp* and *it* refer in this case?

This chapter will offer a practitioner-oriented ontological account of such situations, concerning objects playing the role of components in larger technical systems. In the following, I will first analyse the notion of *artefactual object*, which seems more useful, for practical purposes, than that of *artefact*, being based on compliance to design specifications independently of the artificial origin of the object at hand. I will then focus on *artefactual systems*, composed of several *system components* playing different roles, arguing that the way people deal with them presupposes a non-standard ontological behaviour, allowing for complete replacement and virtual presence.

More specifically, I will explore the main choices of a cognitively biased ontology able to account for the different things that engineers and technicians *have in mind* when they speak of artefacts and their components in a technical discourse, independently of any attempt to capture the objective reality out there, whatever this is. This means that we are squarely in the realm of what Strawson (1959) called *descriptive metaphysics*, which 'is content to describe the actual structure of our thought about the world'. I judge this approach as the most productive for our purposes, given that computational ontologies are essentially aimed at facilitating mutual understanding of each other's *assumptions* about reality.

By its very nature, descriptive metaphysics takes a liberal view concerning the introduction of new ontological categories, as long as they are motivated by cognitive distinctions, often reflected by the surface structure of natural language.

In the framework of the present discussion, a crucial question, much debated in the philosophical literature and especially in the first part of the present volume, is whether or not artefacts do have their own ontological status, corresponding to a separate category, in addition to ordinary natural objects, amounts of matter, events, qualities, and so on. Under the applied ontology perspective, this issue has been discussed in detail in Vieu et al. (2008) and Borgo and Vieu (2009), who are in favour of a radical multiplicative approach inspired by Baker's (2004) constitution view of persons and artefacts. They claim that as soon as a physical object is intentionally selected for a particular purpose, a new entity (belonging to the artefact kind) comes into being, which is co-located with and *constituted by* the original physical object. So, according to this view, when I pick up a pebble from the beach to be used as a paperweight on my desk, a new entity comes into existence, which is 'over and above' the pebble itself. Just because of the intentional selection, I have now a new object, a paperweight, which in turn is constituted by a pebble.

In the following, I will adopt a stronger criterion to justify the emergence of a new entity, based on intrinsic, observable properties instead of a merely historical one such as intentional selection. Indeed, I find the emphasis given to intentional selection difficult to defend from a cognitive point of view. For sure, it is hard to think of it as a sufficient condition for artefactuality: is a piece of wood collected for a fire or a stone collected for a stonewall an artefact? Moreover, even the necessity of such a condition seems to be problematic, since the way we talk of artefacts in everyday language, and in particular of their individuation and persistence criteria, seems not so much bound to their *origin* (e.g. being selected for a certain intended function), but just to their actual *features*. I agree therefore with Thomasson (2013) on the fact that focusing on artefacts' features (rather than on intended functions) provides us a better account of their common understanding. I will make however a more radical move, getting rid not only of intended functions but also of intentional creation. This means that my category of interest will not include just proper artefacts, which by definition are intentionally made, but more in general what I call *artefactual objects* which I will assume to be either natural or artificial.

I will take as an essential property of artefactual objects the fact that their features *fit a pre-existing design*. So artefactual objects are *realised designs*, independently whether such realisation is due to nature or not. In the paperweight example, we may suppose that there exists a design (a private or a public one) involving a wooden table with a nice pebble on it, in a handy position close to the upper right-hand corner, to be used as a paperweight. This is a specific *desk arrangement* design specification, which involves a designed object (the table) and an ordinary, non-designed thing playing a functional role. So there is no design specification for the pebble itself: indeed, the creativity aspect of this specific desk arrangement design is exactly to reuse an ordinary thing as a paperweight. Now let's imagine that this design is *partially* realised at my home: the table is there, I know where the paperweight is supposed to be, but its position is empty. I go to the beach, look around, pick up a nice pebble, and soon I discard it since I found a better one. According to Borgo and Vieu, it seems there are now two paperweights (in addition to the pebbles that constitute them), the one I have in my hand and the one I just discarded. I claim there are no new objects: there are just two pebbles, among the many others on the beach, one of which is about to be *used* to realise my design[1] but does not realise any design in itself, simply because there is no design for *it*. My table, on the contrary, is clearly the realisation of a design specification, since it fits a certain description, say, in the Ikea catalogue. I claim that this is an essential feature the table has that distinguishes it from the amount of matter it is constituted by. So, both the pebble and the table are physical objects; only the table, however, is an artefactual object.

In conclusion, let me remark that despite the arguments above against the view that a mere act of selection is powerful enough to create a new ontological entity, intentional selection plays of course a fundamental role in any theory of artefacts and artefactuality, and it is very important to understand its ontological implications. As we shall see, this is not only a philosophical problem, since realisation and maintenance of technical assemblies presuppose a process of intentional selection, composition and replacement of their components. In the following, after an analysis of the ontological status of artefactual objects in the light of the *design realisation* view outlined above, I will explore the ontological implications of intentional selection of components in the course of the *realisation, use and maintenance process* of technical systems, with special emphasis on the problem of replacement of system components such as car headlamps.

11.2 Artefactual Objects

In the past, much effort in philosophy has been devoted to the analysis of the notion of artefact. From a metaphysical point of view, a relevant question is, whenever the key definitory aspects of artefacts have been fixed, whether they should form a genuine ontological category, i.e. a *substantial sortal*, in Wiggins' terms, with its own identity criteria. Now, several definitions have been proposed for 'artefact', including those discussed in the present book, and it is not my intention to discuss them. The minimal definitory condition I will take, which is shared (in the form a necessary condition) by most positions, is that 'an object is an artefact if and only if it has an author' (Hilpinen 2004). If understood in this way, my opinion is that the property *being an artefact* has a debatable ontological status and is probably not relevant from the perspective of applied ontology. Of course, deciding whether or not such a property holds for a particular object (say, for an archaeological find) may be a question of relevant practical and philosophical importance, but this is not enough to justify the choice of considering *artefacts* a genuine ontological category. As I have argued in various papers (Guarino et al. 1994; Guarino and Welty 2000; Guarino 2009), two conditions need to be satisfied for a property to single out a genuine ontological category, i.e. a *substantial sortal*, in Wiggins' terms, or as a *type*, in my own terminology: (1) the property must be *rigid*, that is, essential to all its instances, and (2) the property must supply some common individuation and

[1] Interestingly, only the pebble I have in my hand can be replaced.

persistence conditions for its instances. In my opinion, none of these conditions are satisfied for the property of *being an artefact*.

Concerning the rigidity condition (1), Hilpinen (2004) observes that '[a]rtefact sortals can be essentially or non-essentially (accidentally) artefactual'. The terminology used in this statement is a bit puzzling, but the idea is clear: there are some sortals denoting things which we tend to assimilate to artefacts, like a path through a forest or a word in a language, for which having or not an identifiable author is just an accidental property. This is enough to conclude that the property of *having an author* (and therefore *being an artefact*, under our minimal definition) is not rigid.[2] In other words, at least for certain artefacts (like a word, a path in the forest, a hole in the ground or a new synthetic molecule), it seems hard to believe that they specifically depend for their very existence on some kind of 'authoring' event (being their intentional modification or just their selection for a specific purpose).[3] In addition, it seems natural to suggest a stronger thesis, which I will not explore further here: *being an artefact*, under our minimal definition, is indeed *anti-rigid* in the terminology of Guarino and Welty (2000), in the sense that it is accidental for all its instances and not just for some of them: whether or not a certain object is the product of an author's action (either intentional or not) is just a contingent property.

Concerning the identity condition (2), we must observe that, in our everyday interaction with artefacts, of course we *do* ascribe individuation and persistence criteria to them: however, such criteria are not provided by the property *being an artefact* (indeed, as observed in Guarino and Welty (2002), a non-rigid property *cannot* provide identity criteria). In conclusion, if we assume that *being an artefact* is anti-rigid, it classifies as a *role property* according to Masolo et al. (2004), since it is *definitionally dependent* (Fine 1995) on *being an author* and does not provide identity criteria.

But where do the ordinary identity criteria we use for artefactual objects come from? As Wiggins (2001) observed, and maintained by Lowe (2013) in the present volume, there are indeed no obvious identity criteria provided by generic artefactual *kinds* like *chair*. However, when we focus on a specific chair belonging to an artefactual *species*, e.g. a certain Ikea product, it is certainly easier to point out essential properties and persistent conditions *bound to specific design features*, which, together with general distinguishability conditions for physical objects (like being different if they occupy different spatial location),[4] allow us to conclude that such

---

[2] By the way, I prefer to call the sortals above *artefactual* sortals just to convey the information that their instances are somehow assimilated to artefacts, although improperly. So I would rewrite Hilpinen's statement as follows: 'Artefactual sortals can be essentially or non-essentially (accidentally) artificial'.

[3] Note that what I am denying here is a *de re* dependence: it is obvious that *de dicto*, whether or not something is an artefact depends on the existence of a previous authorship event, but such event is not necessary for the *de re* existence of the thing itself.

[4] Not vice versa, since I admit that some physical objects may be constituted by other physical objects that share the same location. Think for example of those children's toys that can be configured as a starship or a monster or a gun: when the starship disappears, being transformed in a monster, still the toy is there.

species do have an ontological status, i.e. they are *substantial sortals*.[5] So, the reason why only artefactual species – and not kinds – provide identity criteria is that kinds are defined in purely functional ways (i.e. in relation to the capability of fulfilling a predefined goal, like sitting), while species are defined with respect to intrinsic properties, such as shape or physical structure.[6]

This view is another argument against functional essentialism, in line with Houkes and Vermaas (2010): simply put, functionality is not enough to provide identity criteria. The essence of artefactual objects lies in their *design features*, resulting from design choices encoded in *design specifications*. An artefactual object exists only if some rational design choices have been made, resulting in a specific set of design features. Its essential properties and persistence conditions are bound to a certain *degree of compliance* to such design features, whose actual extent is largely a matter of convention.

Note that although the actual choice of such essential properties and persistence conditions will be often subjective, it seems plausible to assume that, at least in the mind of a certain individual, such choices will be common to all the instances of a certain species, while they will not be common to all the instances of a generic artefactual kind. For example, in the case of the artistic exhibit mentioned by Lowe (2013), I agree it would be difficult to devise identity conditions for the 'shedboatshed' kind, but I am convinced that the artist does have his own identity conditions for the particular shedboatshed species (with only one exemplar) he designed.

Coming back to our first example, it should be clear now that if I pick up a pebble to use it as a paperweight, to keep the door open or to build a stonewall, this is not an artefactual object since the stone *species* has not been designed in advance.[7] On the other hand, as Thomasson (2013) notes, if I decide to use a teapot (instead of the pebble) as a paperweight, it will still be recognised as a teapot despite being used as a paperweight, just because it fits the teapot design specifications.

Before continuing, a few clarifications are due. *First*, I have to explain what I mean by design specification. Of course, the standard case is that of engineering blueprints, but I do not require a design specification to be explicitly encoded on a publicly accessible physical substratum: it could also be encoded in the mind of an agent, what is important is that certain choices concerning *how* a certain function will be realised have been made. In other words, the essence of an artefactual object lies not in its mere *capacity* to fulfil a certain function, but rather in the *way* the

---

[5] I arrive therefore at a conclusion similar to Carrara and Vermaas (2009), in the sense that I admit a fine-grained ontology of artefactual types. However, the reason of such fine-grainedness is not due to the fact that objects with the same physical operational principles might have different functions, but simply to the existence of multiple design specifications for the same generic function.

[6] As a methodological note, I believe we should acknowledge a primacy of intrinsic properties (as compared to extrinsic ones) to determine ontological categories.

[7] At least, this is the typical case. Should a particular paperweight species be predesigned in advance, carefully choosing the desired weight and size, then each stone complying with such design specification would constitute a paperweight.

capacity is obtained, thanks to specific intrinsic properties, selected as a result of a rational design process and encoded in a design specification. Note that the *rational choice* of a specific set of *nominal properties* is a key aspect of a design: so a simple constraint such as '*weight* $\geq$*200 g and size* $\leq$*10 cm*' does not count as a design, but just as a *requirement*, i.e. as an input to the actual design.

Second, as should be clear from the previous discussion, such rational choices, resulting from an intentional process, do not concern the artefactual object itself, but just its *species*: therefore, a particular object might happen to comply with a design specification, constituting an artefactual object, without being the result of an intentional selection and/or modification. So, supposing that on Mars there is a stone very similar to what we call a cup on the Earth, such a stone would constitute for us an artefactual object of the *kind* 'cup' if it complies (within a certain tolerance) with one of our design specifications for a cup, while it would be just a strange stone otherwise, possibly being used to play the *role* 'cup', i.e. being just an *instrumental* cup in Dipert's (1993) terminology. On the contrary, if a Martian would find one of our cups, she might consider it just as a natural object, possibly realising after some investigation that it is an artefact, and possibly adding a new entity in her repertoire of artefactual objects only after having assimilated its design specification, corresponding to a new artefactual species.

Third, a design specification does not need to be immutable and complete in all its details before singling out artefactual objects. The design of a house (or of a temporary shelter) can evolve during the construction, similarly to the way works of art do.

In conclusion, let me sketch the structure of the overall taxonomy that I have in mind (Fig. 11.1). I define artefactual objects as physical objects that necessarily comply with some design specifications. The generic kind *chair* includes all physical objects that necessarily comply with design specifications intended to achieve a certain intended function, namely, to let people sit. Note that the degree of such compliance depends on the specification itself, and in general does not require proper functioning: a broken chair is still a chair, up to a certain extent. This leaves open the possibility of malfunctioning artefactual objects.

Note also that, while I subscribe to the view that physical objects are constituted by amounts of matter, I am not saying that, in turn, artefactual objects are constituted by physical objects. That is, I am not extending to artefactual objects the position proposed by Borgo and Vieu (2009), according to which artefacts are a genuine new category in addition to physical objects, being constituted by them. This multiplicative move would be unavoidable if we would maintain that the essential properties of artefactual objects are incompatible with those of physical objects. This is not the case, in my view, since I am assuming just a minimal essential property common to all physical objects, namely, some weak form of maximal unity, admitting the possibility of parts replacement and change of shape and structure. In this way, artefactual species just add further essential properties (such as those related to shape and structure) which do not conflict with these minimal ones, without requiring

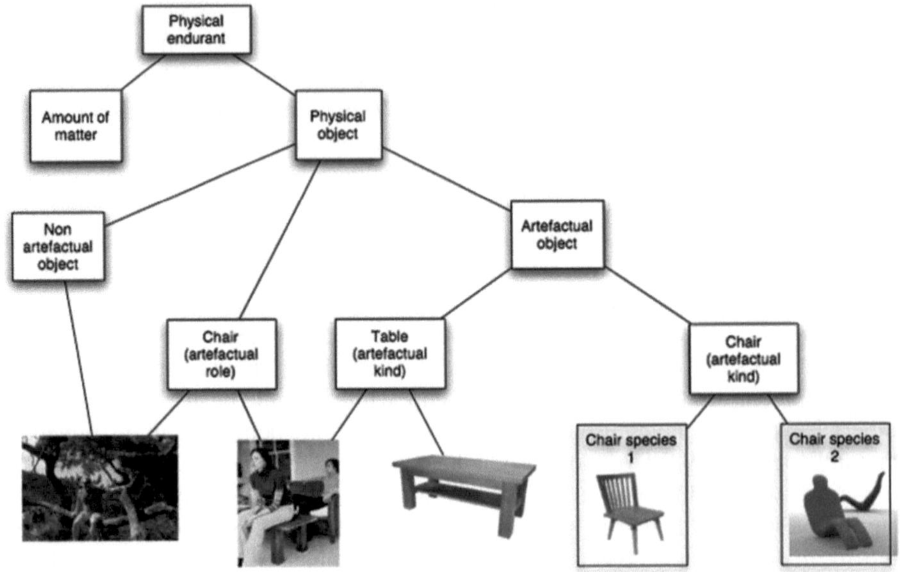

Fig. 11.1 Artefactual objects, artefactual kinds and artefactual roles

therefore the emergence of a new entity.[8] Note however that I do not exclude the possibility that a physical object is constituted by another physical object: indeed, this is often the case (think of a hook made with a piece of metallic wire).

I believe that this position is much simpler to accept from a cognitive and linguistic point of view: maybe it is just a terminological issue, but it would be hard to convince somebody that a chair is not a physical object. Of course, should we adopt a stricter notion of physical object, assimilating them to isolated *pieces of matter* not admitting change of shape and parts replacement, then we would be forced to assume artefactual objects as being different from physical objects, being constituted by them. Indeed, in our case, we could explicitly admit the category of pieces of matter as a specialisation of physical endurants in Fig. 11.1, having physical objects constituted by pieces of matter, and then, in turn, by amounts of matter. This however would require a careful analysis of spatially noncontinuous physical objects like assemblies or scattered objects, which is beyond the scope of the present chapter.

So, in conclusion, in addition to the minimal essential properties of physical objects, we ascribe to artefactual objects some extra essential properties bound to their shape and structure that fit (or in general *comply with*, up to a certain extent) certain design specifications. We say in this case that an artefactual object *realises* a

[8] Those who take intentional selection as an essential property of artefacts cannot exploit this minimal view of physical objects to simplify their theory, since intentional selection is clearly contingent for all physical objects, hence the contradiction, which motivates the choice to assume artefacts as constituted by physical objects.

design specification. Design specifications are developed with the intent that their realisations will have the capability to perform a certain function, but this capability is not intentionally ascribed to all objects that happen to comply with a design specification: it may happen that the capability is not there because the object is malfunctioning (so the compliance is not perfect) or just because the design is wrong, or perhaps the capability is there, but it is not ascribed to the object by anybody, because that amount of matter with such specific shape and structure has a natural origin, and nobody selected it on the basis of its capability (that last amount of matter still counts as an *artefactual* object, but it is not an artefact).

Under this view, as we discussed for the cup example, a natural object like the trunk in Fig. 11.1, which just happens to be used as a chair, is not an artefactual object belonging to the kind 'chair', since there is no design specification for it: it just *plays the role* of a chair. So, for a lexical item like *chair* we have two senses: a *proper* one, denoting an *artefactual kind*, and an *extended* one, denoting an *artefactual role*, including all objects that are used as chairs.[9] According to Masolo et al. (2004), the property corresponding to the former sense is a *rigid* one (i.e. essential for all its instances) while the property corresponding to the latter is *anti-rigid* (i.e. contingent for all its instances) and *definitionally dependent* in the sense that defining what a chair is requires the introduction of other properties holding for other things (namely, for people).

11.3 Artefactual Systems and System Components

Let us now explore more in detail the way a relevant kind of artefactual objects, namely, *artefactual systems*, are realised and maintained – on the basis of their design specifications – by means of intentional selection, composition and replacement of their components. Typically such components are artefactual objects themselves, like in the case of *technical assemblies*, but this is not necessarily the case, as in the paperweight example, since sometimes the design of an artefactual system can include natural objects as components.

We have seen that the intentional selection of an object as a *candidate* functional component for a larger system does not justify the creation of a new entity, so, let us say, it is not enough to modify the world. A different case, however, is when the selected component is put in place, possibly replacing a previous one. To analyse the ontological assumptions behind these situations, the idea of *direction of fit* discussed (although in a different context) in the philosophical literature since Anscombe (1957) might be of help. Let us assume I have a certain design specification of an artefactual system in my mind (say, the desk arrangement described above). Now suppose I want to create a concrete *realisation* of such specification.

[9] Note that the extension of such artefactual roles intersects with the extension of the corresponding kind: the things being actually used as a chair certainly include some artefactual chairs, but not all of them: think of chairs that are never sold and used, and scrapped afterwards.

When I select a specific object, say, a nice wooden table I found in a shop, as a candidate for being a component of my system, a specific mental representation of such system is created, which more or less complies with the design specification but *adapts* itself to the particular object I selected. If, on the way home, I find a nicer table, I may decide to use the previous one for another purpose, selecting the new one for my desk. In this case, the mental representation of my desk changes, but the world has not changed yet to fit my plans: this only happens when I finally arrive at home, choose a particular place and put the table there. The desk I designed is now in place, but *the paperweight is missing*. Note that *the paperweight* seems to denote now something that shares some characteristics of physical objects: in particular, it has now its own position (namely, in the right-hand corner of my new table), although nothing is there yet. I name *it*, and I have *it* in mind when I go to the beach to look for the right pebble whose colour matches that of my table. When I will finally pick up a pebble among many other candidates, the world will not change just because of that selection, but it did change when I put the table in place, starting realising my working desk and will change again when I will put the pebble in the paperweight's place.

So, when I select a component to fit an artefactual system's design specification, it is the mental representation which gets changed, not the world, because the representation now includes that object and not another one (say, a white pebble instead of a black one). On the contrary, when I select a certain physical place to start realising the structure I have in mind or an existing partially realised artefactual object to complete or modify it, it is the world that will be changed. In other words, in the *first* direction of fit, the mind comes closer to the world: the object in the world, once selected, draws my design specification in its direction, so to speak, since the specification adapts to it. In the *second* direction of fit, it is the world that moves closer to the mind and adapts to it in order to fit my intention. As I see it, the hard ontological problems that may require a radically multiplicative (and possibly non-standard) approach, as discussed below, concern *the latter* direction of fit, not the former one, since we need a way to refer to the new entities created by a design realisation process. So we don't need to admit the emergence of new individuals in an ontology of artefactual systems during the *component selection* process, but definitely we need to consider the ontological implications of the *component (re)placement* process.

Focusing now on our initial puzzle, let us consider the way we refer to components of artefactual systems (say, my car's right headlamp) in ordinary discourse. Consider, for instance, the case where the headlamp is being replaced. When the lamp is removed, still a technician can say 'this cable goes to the right headlamp'. To what does he refer? And what about the statement 'the right headlamp has been replaced twice'? What does 'my car's right headlamp' denote?

My answer is that, in all the cases above, we refer to a special imaginary, conventional entity, which is specifically dependent on both a physical object (my car) and a design specification. Such an entity has two modes of presence in time and space: it is *virtually present* at a certain physical location (contiguous to the headlamp socket) when no lamp is there and is *actually present* when a lamp is fitted into the

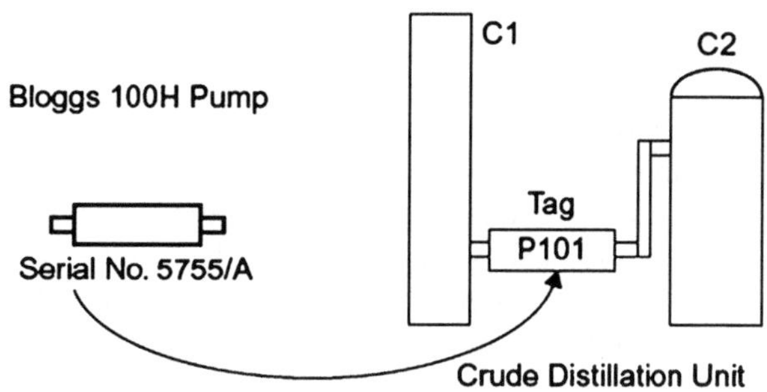

Fig. 11.2 Conventional system components and ordinary physical objects (Courtesy from West (2010, p. 176))

socket. This is the way this conventional entity maintains its identity during its life cycle, which might involve multiple lamp replacements.[10]

Trying to stay close to engineering practice and adapting an expression used in West (2010), I will call such entities *conventional system components* (or simply *system components* where the context will be clear).[11] Although it seems hard to choose a suitable common name for these things, for sure engineers use proper name tags for them. For example, Fig. 11.2 shows a picture taken from West (2010), where the tag 'P101' is used to denote a pump playing a specific role in an oil distillation unit. This tag is typically annotated in design specifications and replicated for specific product realisations together with the product identification number. Of course, as we can see from the figure, the system component's identifier is very different from the serial number of specific physical parts, such as the particular pump being installed at a certain time.

Let us now investigate the ontological status of conventional system components more in detail. First of all, the reason I stress they are *conventional* entities is that they are not physical objects, as it does not seem plausible to maintain that physical objects can exist in a virtual state and survive complete replacement of their constituents. According to DOLCE's upper-level ontology (Masolo et al. 2003), they are however *physical endurants* (or simply *objects* according to Borgo and Masolo (2009)), since at each time they are present they have a spatial location, although they might happen to be immaterial (like a hole, for instance) at some times. Moreover, conventional system components share many properties with artefactual

[10] My intuition is that, when something is replaced, something *else,* namely, a kind of imaginary *place*, is assumed to exist in addition to the object being *replaced* in that role. In this sense, conventional system components can be seen as special places, dependent on their host, where something relevant may happen.

[11] Of course, the abbreviated expression only works as long as we focus on artefactual systems only. If we consider larger systems that include non-designed components (for instance, an airplane including its passengers), such components may behave differently from the system components we are discussing here. See Franssen and Jespersen (2009) for an account of such situations.

objects, as they still satisfy a unity criterion and comply with design specifications, but exhibit some radical differences with respect to them, namely:

1. A conventional system component is *constantly specifically dependent* (Thomasson 1999) for its existence on an artefactual object, called its *host*: the host of the headlamp is the car, and the host of the bulb is the headlamp group.
2. A conventional system component is *actually present* at a certain time if at that time there is a physical object whose position, shape and structure comply with the design specification of the hosting object. Such physical object is called the *physical constituent* of the system component. Note that this definition excludes the possibility to state that the car's headlamp is actually present if, say, a piece of paper is put into the lamp socket.
3. A conventional system component is *virtually present* at a certain time, at the location foreseen by the host's design specifications, if it is not actually present at that time and its host is present at that time. Note that the assumption that conventional system components can be virtually present marks a difference with West's approach.
4. A conventional system component is characterised by a number of *nominal qualities*, whose values and mutual constrains are specified in the host's design specifications.[12] So our headlamp will have a nominal weight, power, shape, and so on. These nominal qualities are different from standard physical qualities, as they inhere to different entities, but share the same quality spaces.[13] When a system component is actually present, it also inherits the qualities of its constituent, so when a lamp different from the recommended one is mounted, we can say that the headlamp has a nominal power of 35 W, while the actual power is, say, 50 W.

So, the first peculiarity of conventional system components compared to artefactual objects (and physical objects in general) is that they can be *empty*, in the sense that they can still exist – although in a virtual status – even when they have no physical constituent. In a sense, conventional system components are like ghosts that can materialise and disappear; however, differently from ghosts, they maintain their spatiotemporal identity, since the *place* they can materialise is determined by the host object and by our expectations concerning the potential filler of such location, based on the host's design specifications. Among other advantages, admitting the existence of conventional system components allows us to account in a simple way for the formal surface semantics of statements like *the right headlamp of my car is missing*, avoiding complex ad hoc paraphrases: *missing* takes as argument a conventional system component and expresses the fact that such component has no physical constituent.

A further peculiarity of conventional system components is that they can be *substituted*, i.e. they can undergo a *complete replacement* of their physical constituent as a whole, yet maintaining their identity. This is typically not admitted for artefactual objects (and physical objects in general), whose parts might perhaps be completely

[12] If the host artefactual object uses *standard* components, their specifications might be separate.

[13] I am adopting DOLCE's account of qualities (Borgo and Masolo 2009).

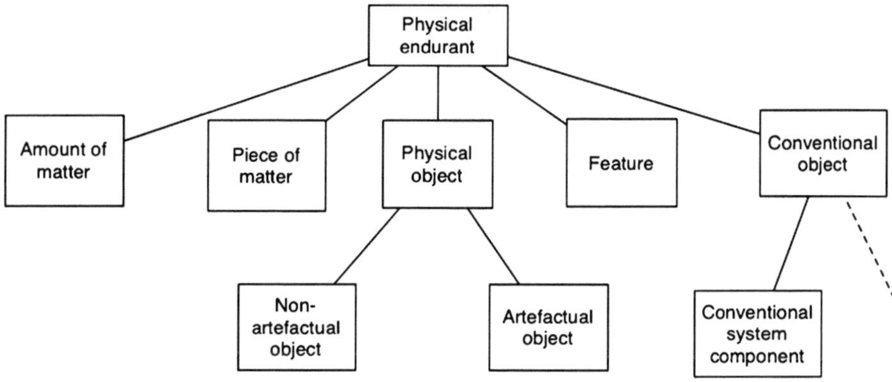

Fig. 11.3 Conventional system components as a kind of physical endurant

replaced only in a continuous way, like in the puzzle of the ship of Theseus.[14] Like in the previous case, admitting conventional system components in our ontology allows us to provide a simple solution for the formal semantics of statements like *the right headlamp of my car has been replaced twice*, which requires an understanding of *what* has been replaced (clearly not the physical lamp). The answer is simple: it is the conventional system component which changed its constituent two times.

Finally, an important difference between conventional system components and artefactual objects is that the former might not have their *own* design specifications, while still complying with the overall design specifications of their hosts, as we have seen for the paperweight example. Another example can be a stonewall made of natural stones: the wall has a design specification, the individual stones do not.

It might be interesting to compare the ontological status of conventional system components to that of *features*, introduced in the DOLCE ontology (Masolo et al. 2003). Features include *relative places* like a hole in the wall or the underneath of the table, as well as relevant parts marked by morphological discontinuities such as the nose of a statue. So features can be either permanently *empty* or permanently *filled* with matter. Conventional system components, on the contrary, can be filled at a certain time and empty at another time. Abstracting from artefactual objects, we can generalise this behaviour by introducing the general category of *conventional objects*. A river or a lake getting dry in summer and filling up in winter, with most of their water being constantly replaced, would belong to this category, as well as perhaps an African tribal spirit, which needs a mask to mark its presence independently of the person who wears the mask and is virtually present in the mask even when nobody wears it. Along these lines, we might revisit the ontology of social objects under a constructivist stance, but this is clearly beyond the scope of the present chapter. Anyway, Fig. 11.3 gives an idea of the ontology of physical endurants I have in mind.

[14] I adopt a very opportunistic approach concerning the Ship of Theseus problem: the rules for deciding whether an artefactual object survives certain parts replacement depend on the specific artefactual species and are ultimately a matter of social conventions.

11.4 System Components, Replaceability and Functional Roles

After this analysis of artefactual systems and their conventional components, let us take a step back, trying to understand the *raison d'être* of conventional system components from a broader perspective. As they are realisations of design specifications, we can say that system components play a *functional role* within the artefactual object. In the applied ontology literature, functional roles are considered to be a special case of *social roles* (Loebe 2007), implying some kind of agent's expectations on the player's behaviour (independently whether such behaviour pertains to another agent or an inanimate object). Now, the present analysis shows that functional roles exhibit a *peculiar* ontological behaviour, which is not shared by all social roles nor by other kinds of role. Briefly said, their players can be *replaced*, instead of just *changed* like for ordinary roles. Consider, for instance, a truck transporting a certain load. We can replace a wheel, a lamp or even the driver, but we don't usually say that the truck load is replaced, it is just a new load. The difference between *replacement* and *change* of a role player is definitely subtle, but it seems to be related to the fact that the possibility of replacement implies that something can maintain its identity despite the replacement. Now, it looks like only functional roles admit such replacement possibility for their players. So, in the light of the above discussion, my conjecture is that *for all functional roles*, but not for other role kinds, we need to admit the existence of an idealised, conventional role player, whose physical realisation can be replaced depending on the circumstances. Therefore, the picture I described in this chapter seems not to be limited just to artefactual objects but appears to have a more general relevance, being bounded to the general notion of functionality independently of artefactuality.

For sure, anyway, admitting conventional components in our ontology seems to be a definite advantage for a common-sense account of the engineering domain: we have seen how in many cases of ordinary engineering language we refer to the conventional role player, abstracting from the actual physical player; in addition, even when we refer to the physical player, we need to distinguish between the actual behaviour and the expected behaviour, this is why the distinction between actual qualities and nominal qualities turns out to be crucial.

In conclusion, coming back to the original goals of this chapter, let us consider the following statements, taken from ordinary technical language, and see how the ontological analysis discussed above can help clarify and simplify their formal semantics:

1. The right headlamp has been smashed in an accident.
2. The right headlamp is missing/uninstalled.
3. The right headlamp has been replaced twice since I bought it.
4. This cable leads to the right headlamp.
5. The right headlamp has a nominal power of 35 W, but the lamp I mounted has a power of 50 W.

In (1), clearly '*the right headlamp*' refers to the physical component, since a conventional component cannot get smashed. In (2), on the contrary, the only possible choice – if we want to maintain the sentence's surface structure – is to refer to the conventional component, as we discussed. The same holds for (3): in this case, the conventional component is the common participant to the various replacement events, each involving a different physical component. In (4), there is probably some ambiguity, but if we want the superficial structure of this statement still makes sense when the right headlamp is missing, then the only alternative is to assume a reference to the conventional component. Finally, (5) shows a case where we have to reason with nominal qualities ascribed to conventional components, comparing them with the actual qualities of physical components.

Acknowledgments This work has been performed in the framework of the European Marie Curie International Research Staff Exchange project EUJOINT. I would like to thank Stefano Borgo, Emanuele Bottazzi, Maarten Franssen, Yoshinobu Kitamura, Peter Kroes, Riichiro Mizoguchi, Pieter Vermaas and Laure Vieu for their thoughtful comments, their suggestions and the lively discussions. I would like to thank also my partner Rosalba for her patience, understanding and support.

References

Anscombe, G. E. M. (1957). *Intention*. Oxford: Basil Blackwell.

Baker, L. (2004). The ontology of artifacts. *Philosophical Explorations, 7*, 99–111.

Borgo, S., & Masolo, C. (2009). Foundational choices in DOLCE. In *Handbook of ontologies* (2nd ed., pp. 361–381). Berlin: Springer.

Borgo, S., & Vieu, L. (2009). Artefacts in formal ontology. In A. Meijers (Ed.), *Handbook of philosophy of technology and engineering sciences* (pp. 273–308). Amsterdam: Elsevier.

Carrara, M., & Vermaas, P. E. (2009). The fine-grained metaphysics of artifactual and biological functional kinds. *Synthese, 169*, 125–143.

Dipert, R. (1993). *Artifacts, artworks, and agency*. Philadelphia: Temple University Press.

Fine, K. (1995). Ontological dependence. *Proceedings of the Aristotelian Society, 95*, 269–290.

Franssen, M., & Jespersen, B. (2009, June 15–17). From nutcracking to assisted driving: Stratified instrumental systems and the modeling of complexity. In *Second international symposium on engineering systems*. MIT, Cambridge, MA.

Guarino, N. (1998). Formal ontology in information systems. In N. Guarino (Ed.), *Formal ontology in information systems* (pp. 3–15). Amsterdam: Ios Press.

Guarino, N. (2009). The ontological level: Revisiting 30 years of knowledge representation. In A. Borgida, V. Chaudhri, P. Giorgini, & E. Yu (Eds.), *Conceptual modelling: Foundations and applications, essays in honor of John Mylopoulos* (pp. 52–67). Berlin: Springer.

Guarino, N., & Musen, M. A. (2005). Applied ontology: Focusing on content. *Applied Ontology, 1*(1), 1–5.

Guarino, N., & Welty, C. (2000). A formal ontology of properties. In R. Dieng & O. Corby (Eds.), *Knowledge engineering and knowledge management: Methods, models and tools, 12th international conference, EKAW 2000* (pp. 97–112). Juan-les-Pins: Springer.

Guarino, N., & Welty, C. (2002). Identity and subsumption. In R. Green, C. Bean, & S. Myaeng (Eds.), *The semantics of relationships: An interdisciplinary perspective* (pp. 111–125). Dordrecht: Kluwer.

Guarino, N., Carrara, M., & Giaretta, P. (1994). An ontology of meta-level categories. In J. Doyle, E. Sandewall, & P. Torasso (Eds.), *Principles of knowledge representation and reasoning: Proceedings of the fourth international conference (KR94)* (pp. 270–280). San Mateo: Morgan Kaufmann.

Hilpinen, R. (2004). Artifact. In E. N. Zalta, U. Nodelman, & C. Allen (Eds.), *Stanford encyclopedia of philosophy*. Stanford: Center for the Study of Language and Information, Stanford University.

Houkes, W., & Vermaas, P. E. (2010). *Technical functions: On the use and design of artefacts*. Dordrecht: Springer.

Loebe, F. (2007). Abstract vs. social roles: Towards a general theoretical account of roles. *Applied Ontology, 2*, 127–158.

Lowe, J. (2013). How real are artefacts and artefacts kinds? In M. Franssen, P. Kroes, T. A. C. Reydon, & P. E. Vermaas (Eds.), *Artefact kinds: Ontology and the human-made world* (pp. 17–26). Heidelberg: Springer.

Masolo, C., Borgo, S., Gangemi, A., Guarino, N., & Oltramari, A. (2003). *Ontology library: The DOLCE ontology. WonderWeb deliverable D18*. Available online at wonderweb.semanticweb.org/deliverables/documents/D18.pdf

Masolo, C., Vieu, L., Bottazzi, E., Catenacci, C., Ferrario, C., Gangemi, A., & Guarino, N. (2004). Social roles and their descriptions. In *9th international conference on principles of knowledge representation and reasoning (KR 2004)* (pp. 267–277). Palo Alto, CA: AAAI Press.

Munn, K., & Smith, B. (2010). *Applied ontology: An introduction*. Frankfurt: Ontos Verlag.

Strawson, P. F. (1959). *Individuals: An essay in descriptive metaphysics*. London: Routledge.

Thomasson, A. L. (1999). *Fiction and metaphysics*. Cambridge: Cambridge University Press.

Thomasson, A. L. (2013). Public artefacts, intentions, and norms. In M. Franssen, P. Kroes, T. A. C. Reydon, & P. E. Vermaas (Eds.), *Artefact kinds: Ontology and the human-made world* (pp. 45–62). Heidelberg: Springer.

Vieu, L., Borgo, S., & Masolo, C. (2008). Artefacts and roles: Modeling strategies in a multiplicative ontology. In C. Eschenbach & M. Gruninger (Eds.), *Proceedings of the 5th international conference on Formal Ontology in Information Systems* (*FOIS08*) (pp. 121–134). Saarbrücken: IOS Press.

West, M. (2003, September 23). Replaceable parts: A four dimensional analysis. In *COSIT-03-workshop on fundamental issues in spatial and geographic ontologies*. Ittingen.

West, M. (2010). *Developing high quality data models*. Burlington: Morgan Kaufmann.

Wiggins, D. (2001). *Sameness and substance renewed*. Cambridge: Cambridge University Press.

Chapter 12
Engineering Differences Between Natural, Social, and Artificial Kinds

Eric T. Kerr

Abstract My starting point is that discussions in philosophy about the ontology of technical artifacts ought to be informed by classificatory practices in engineering. Hence, the heuristic value of the natural-artificial distinction in engineering counts against arguments which favour abandoning the distinction in metaphysics. In this chapter, I present the philosophical equipment needed to analyse classificatory practices and then present a case study of engineering practice using these theoretical tools. More in particular, I make use of the Collectivist Account of Technical Artifacts (CAT) according to which there are different classificatory practices for natural, artificial, and social objects. I demonstrate that in the community studied, artificial kinds are marked by distinctive classificatory practices. The presence of these distinctive classificatory practices in engineering with regard to artificial kinds should inform discussions about the ontology of technical artifacts just as the distinctive classificatory practices in natural science inform discussions about natural kinds.

Keywords Kinds • Ontology • Philosophy of engineering • Strong Programme • Technical artefacts

---

E.T. Kerr (✉)
Science, Technology, and Society Cluster Asia Research Institute,
National University of Singapore, Bukit Timah Campus,
469A Tower Block #10-01, Bukit Timah Road, Singapore 259770
e-mail: e.t.kerr@sms.ed.ac.uk

M. Franssen et al. (eds.), *Artefact Kinds: Ontology and the Human-Made World*,
Synthese Library 365, DOI 10.1007/978-3-319-00801-1_12,

12.1 Introduction: The Continuity of Philosophy, Science, and Engineering

Here is a fairly typical statement from the philosophical literature on natural kinds:

> Scientific disciplines divide the particulars they study into *kinds* and theorize about those kinds. To say that a kind is *natural* is to say that it corresponds to a grouping or ordering that does not depend on humans. We tend to assume that science is successful in revealing these kinds; it is a corollary of scientific realism that when all goes well the classifications and taxonomies employed by science correspond to the real kinds in nature. (Bird and Tobin 2008)

Different scientific disciplines may be accorded different privileges with respect to dictating what kinds are real. Chemistry may be taken to supply paradigmatic examples of natural kinds – chemical elements, compounds, and so on. Physics and astronomy may be second in line providing such kinds as electrons, quarks, and red dwarves. Taxonomies in biology were once taken to be paradigmatic but are now generally thought not to represent natural kinds.[1] Nevertheless, it would be inconceivable that, were science to discover a fault in its own classificatory systems, this would not force a change in philosophical ontology. It is inconceivable, for example, that metaphysicians would continue their 'essentialist' talk about species following the advent of Darwin's theory of evolution (Dupré 1993, p. 3; Okasha 2002, p. 191). Similarly, quantum physics forced changes in certain metaphysical theories such as determinism (de Koninck 2008), and the philosophy of space and time has been brought through paradigm shifts with Newton, Mach, and then Einstein (Whitrow 1980). The continuity of philosophy and science – the idea that philosophy ought to heed the results of empirical investigation in the natural sciences – has become orthodox, albeit a controversial orthodoxy, in contemporary analytic philosophy of science. Following this line of thought, philosophers of technology might say that if the natural sciences should inform the ontology of natural kinds, the engineering sciences should inform the ontology of artificial kinds.[2] That is to say, the privilege which scientists have enjoyed when it comes to classifying what Mario Bunge (1977), David Bloor (2007), and the authors of the introduction to this volume have metaphorically called the 'furniture of the world' (i.e. its basic constituents, what classificatory systems refer to) should also be accorded to engineers when it comes to classifying the actual *furniture* of the world – its made, or otherwise artificial, objects. If the reader accepts this thesis, then the purposes of this chapter should be clear: to ascertain what kinds of entities engineers do posit with a view to including them in our ontological framework.

Despite what I have said above, that there is a distinction at all between natural and artificial things has recently been called into question (Baker 2008; Pitt 2011;

[1] Seminal works by Kripke (1980) and Putnam (1975) refer to biological kinds as natural kinds although more recent work has challenged this (Dupré 1993; Sober 1994, p. 163).

[2] The view that engineers are in a position of authority with respect to the nature and our knowledge of artificial kinds (or, alternatively, technical artifacts) has been proffered by Garbacz (2012), Layton (1974), and Vincenti (1990).

Preston 2008). Others have argued for the stronger position that not only are artificial things not significantly distinct from natural things but they do not really exist at all (Hoffman and Rosenkrantz 1997, p. 173). Lynne Rudder Baker cites digital organisms, robo-rats, bacterial batteries, and 'search-and-destroy' viruses as technological advances that blur the ontological distinction between natural and artificial kinds (Baker 2008, p. 2). As has been pointed out elsewhere, the blurred line between the natural and the artificial is not a consequence of modern technology, as these state-of-the-art examples might indicate, but has been present for at least as long as technology itself: as long as, for instance, the domestication of cattle or the agricultural cultivation of land and crops (Preston 2008). Synthetically created chemical compounds, genetically modified foodstuffs, and many other things could cause us problems if we were to draw a fine line between natural and artificial things. What this suggests is that the distinction between the natural and the artificial has been problematic for a long time. Baker predicts that the distinction that is drawn between natural and artificial kinds on the basis of their mind-dependence or mind-independence (that is, in some sense, that the former do not depend on human minds whereas the latter invariably do) will continue to be eroded by future technological advances (Baker 2004, p. 15, 2008, p. 5).

One normative constraint that is placed on putative distinctions which is often elided by philosophers interested in formal definitions is that they are useful. In other words, classifications are, on this understanding, interest relative (Searle 2002, p. 59). We carve nature at various joints to suit our purposes of inquiry, knowledge expansion, and, in the case of engineering, to reshape our environment to better suit our needs. They are also profession relative. That is, as Ian Hacking puts it, 'some kinds are essential to some crafts' (Hacking 1991, p. 123). Hacking's argument is that we should not expect one particular taxa, which has been formed historically to suit particular theoretical and experimental needs in one subject, to also be an excellent model for other fields of inquiry. None of this should be read as implying that the physics taxa or the biological taxa are the real taxa whilst the engineering taxa are not of, in Baker's terms, 'genuine substances' (Baker 2008, pp. 3–4). Rather, we place certain authority on expert classifiers, and these professional taxonomies ought to be continuous with or at least inform philosophical ontology.

In the following section, I will describe a theory of classificatory practices that is particularly suited to empirically analyse reference practices to technical artifacts within the engineering professions, and the results of which should be taken into account in discussions of the ontology of technical artifacts.

12.2 Outlining a General Methodology for the Collectivist Account of Technical Artifacts

The Collectivist Account of Technical Artifacts (CAT) is not so much concerned with the formal structure and definition of technical artifacts as with the referencing and classificatory practices about them. It adopts an ontological framework developed by Martin Kusch, following the sociologist Barry Barnes (Barnes 1983; cf. Kusch 1999; Schyfter 2009). Reference is a protean word, but we can define it simply as that which connects 'our speech and that which is spoken of' (Barnes 1983, p. 524). More generally, reference is the kind of behaviour humans exhibit when they take an object *qua* object of a particular kind. That is, when they speak, for example, of water as H_2O, or when they treat it as a natural kind, or when they reinforce another's behaviour taking it to be a particular kind, or when they sanction perceived incorrect behaviour taking it to be a member of another kind of thing. For instance, when I say that this chemical compound is a vitamin and is required in the human diet, I take the chemical compound to be a natural kind, a kind of thing whose description is exhausted by physical-chemical description. Martin Kusch delineated three different kinds of reference that we may use when taking an object to be a particular kind of thing.

To provide a brief summary, consider what would happen to vitamins if we were to cease to use the word 'vitamin' to refer to certain objects as we currently do. The answer is that those objects to which we usually referred when we said 'this is a vitamin' will continue to exist. All that has changed are our classificatory practices. Doubtless this often occurs when we re-examine, in the light of scientific evidence, what objects precisely our terms do refer to. These objects exist, as it were, independently of our referencing practices. Kusch calls these terms, whose description is exhausted by physical description, natural kind terms.[3] On the other hand, consider another group which includes things like marriage, money, and philosophy professor. The meaning of these terms is entirely exhausted by our referencing practices. If a nation decides to change currency, for example, the old currency – the bits of paper or metal coins or whatever they might be – do not retain their ontological status. In fact, they themselves are fundamentally changed by changes in our referencing practices. If we, as an economic community, ceased to treat British pounds as British pounds – if all the institutional frameworks that give those bits of paper, metal discs, plastic cards, and electronic data its status and if all the citizens in that community stopped taking those objects to be British pounds – then there simply would be no British pounds any more. They would pop out of existence in a way that vitamins would not if all the science departments and pharmaceutical companies closed. Kusch calls these terms, whose description is exhausted by social, intentional description, social kind terms.

Kusch notes that there is a third kind term, which we may call artificial kind terms. Artificial kind terms rely for their meaning on both natural and social reference practices (see also Schyfter (2009)). This chapter is an attempt to apply this analysis of three different kinds of reference practices to a specific case, namely, petroleum engineering. In particular it analyses reference practices with regard

[3] I have switched here to talking about natural and social kind *terms* as opposed to natural and social kinds to stay in step with Kusch's account. The distinction trades on a methodological commitment to analysing the words as they are used rather than metaphysical rumination upon the thing-in-itself.

to the artificial kind term 'tool'. The case study was set up in the following way. First, in communication with several engineers, I discussed what tools are (see below) and developed five kinds of reference that might be made. Five documents from petroleum engineering were then analysed for mentions of 'tool' or 'tools', and the kind of reference was noted. I then look at the most dominant kinds of reference made and extrapolate from this a definition of 'tool reference' in the community. It appears that this practice of tool reference is not reducible to either natural or social kind terms but has its own patterns of reference. Since referencing practices in the natural sciences have given ontological status to such entities as electrons, quarks, and galaxies, we can deduce from this that ontological status ought to be conferred upon such entities as tools. It will be for future research projects to analyse referencing practices and identify other artificial kinds that may inform the philosophical ontology of technical artifacts. CAT does not argue that these objects are real but rather suggests that one empirical entry point for understanding what kinds of objects are real is through classificatory and referencing practices in professional disciplines. Before we can apply the CAT approach to the case study, however, it will be necessary to have a closer look at the distinction between natural, social, and artificial terms.

12.3 The (N/S/A) Kind Terms Distinction

12.3.1 Quartz and Quarks: Natural Kind Terms

Before we turn to our empirical analysis, let us discuss in greater detail the distinction between natural, social, and artificial kinds. We tend to think of most things we encounter as belonging to kinds: things which may be treated or taken as being similar to other things of the same kind. This orders our conceptualization of the world, makes sense of our environment, and guides our behaviour. Indeed, it may be said that if we could not identify any object as belonging to a set of things with which it shares certain qualities, exhibits certain regular behaviour, and so on, we would find it extremely difficult to undergo the variety of tasks we wish to accomplish. We can group elements in the Periodic Table or construct a taxonomy of living beings, and yet in these cases, we have come across a classificatory system, in some sense, already in place before we got there or, perhaps, that we were naturally, necessarily, or inevitably inclined towards. It is as if we could not help but group those objects together in the way that we do and to do otherwise would seem awkward. Of course, this is an oversimplification. Our classifications of elements in physics or biological taxa are subject to historical change. Nevertheless, there is a feeling that we do get it right most of the time and if humans were, alas, to disappear from this Earth, we would not necessarily take the elements and animals with us. Central to Barnes' account is what he calls a 'performative model'. Barnes claims that, for kinds of things, the reference becomes fixed in our language because each person has a stored pattern which they compare to the object before them and attach a label accordingly. For example, as I approach what I think looks like a particular flower, I expect certain empirical characteristics: petals, a stem, and perhaps a fragrant smell. If what I encounter matches my internally stored pattern for 'flower', then I will (typically) identify it as a flower. If it does not, I will put it in the 'non-flower' category. Either of these confirming or disconfirming references adds to my community's proper use of the term 'flower'. Most of the time, correct or proper use of a term lies in the general disposition within the community. That is, whether or not a member of the community would typically be disposed to judged that use proper. The proper application of 'flower' will likely differ across lay communities, horticulturalists, and so on, and is liable to change if enough agents change their behaviour. The proper application of *'Tulipa gesneriana'*, on the other hand, will change only incrementally and slowly, if at all. In general, however, all these terms are applied by comparing empirical characteristics to a pattern, model, norm, or prototype. The iterative use of a term to refer to objects presenting similar empirical qualities contributes, positively or negatively, to the proper use of a term relative to a community. It is this pattern that is of interest to CAT.

12.3.2 Marriages and Mortgages: Social Kind Terms

The situation cannot be the same for social kind terms. There are no shared empirical characteristics for 'marriage', for example. No superficial pattern recognition could fix proper use. Marriage may be *symbolized* through wedding rings or another physical object, but these are not necessary to constitute a marriage. If a person loses her wedding ring, for example, or burns her marriage certificate, the marriage remains intact or, if you like, we may still properly refer to the couple as 'married'. However, if our community does not institutionally recognize the couple as married, then there is no sense in which they can be said to be properly married. To put the point more starkly, if everyone in our community suffered some form of collective amnesia which removed the term 'marriage' from our vocabulary and all its associated institutional practices, then it would no longer make sense for anyone to group that couple together with other previously married couples as belonging to the same kind; that vocabulary and the institutional practices associated with it precisely are the thing itself. Social kinds are considerably more nebulous and varied in their properties and behaviour than natural kinds, and the variety of social kind term references is a consequence of this. Consensus, concord, and agreement between community references, together with the initial instantiation which 'primes' (Barnes 1983, p. 529) or baptizes (Kripke 1980) the whole process, form this stereotypical account of social kind terms. If the stereotype for referencing natural kinds is broadly ostensive – identifying a thing by comparison with previous examples – then the stereotype for referencing social kinds is broadly stipulative, defining a thing by stating it as such. Marriage, as with other social kinds, is constituted by a series of performative utterances or 'speech acts'; the 'I do' in some marriage ceremonies being a paradigmatic example (Austin 1975, p. 5).[4]

12.3.3 Microscopes and Gyroscopes: Artificial Kind Terms

Barnes' theory of kinds was incomplete. Kusch (1997, 1999) presents an augmented version of Barnes' account that added artificial kind terms to the natural and social. His reason for doing this is that artificial kind terms (terms that refer mostly to made things) are not wholly identified by their empirical characteristics but neither can they be reduced to an entirely social system of fiat (stipulative definition) and consensus. The referent of a social kind term, for Kusch, is entirely constituted by the references themselves. Talk (and behaviour which takes the referent, R, *to be an R)* actually creates the referent and so, in the final analysis, the referent is the talk (and behaviour) itself. 'The reference… is, as it were, 'exhausted' by the self-reference' (Kusch 1999, p. 245). Consequently, if the references cease to be made, the referent simply drops out of the ontological picture. Natural kind terms, on the other hand, have an 'alter-reference'; the term refers away from the talk towards something in the physical world, things that exist independently of the talk. There are still central patterns, models, norms, prototypes, exemplars, and so on, for what constitutes, say, quartz, and so even natural kind terms have a self-referential collectivist component. However, the patterns, etc. is not the thing itself. Artificial kind terms are odd cases in this framework, however stereotypical it may be. 'Gyroscope', for example, does not just refer to talk (and behaviour) about gyroscopes. However, nor can they be properly identified solely through reading off their empirical characteristics. It has an alter-reference to the physical composition and structure of the object. The reference to a gyroscope is partly a reference to a system or community in which the gyroscope has a function and partly a reference to the physical materials that make it up. One might call to mind the aftermath of the knowledge 'cull' imagined by Alasdair MacIntyre in *After Virtue* (MacIntyre 1984). Here, all scientific artifacts and texts are completely or partly destroyed and all that is left of a previous scientific culture (ours) is fragments and remnants. None of the survivors' descendants attempting to piece together the forgotten knowledge understands what the things are *for*. It is this element that is missing in an account that describes artifacts without reference to a community of users. In order to be more than just a collection of materials, a technical artifact requires a performative social institution referring to

[4] Some legal systems contain a principle of 'common-law marriage' or 'marriage by habit or repute'. In these cases, the married couple may be legally recognized as married even though no official marriage ceremony is performed or marriage contract is entered into. This does not count against Barnes' account. We should read 'referencing practices' broadly to include the kinds of behavior that would, *ceteris paribus*, be required for a couple to be properly referred to as married.

the artifact *qua* artifact. Part of the task of collectivist studies is to identify just what 'referring to an artifact *qua* artifact' is: does it involve design, purpose, functionality, instrumentality, normativity, and so on? What do the referencing practices of the community reveal about these questions?

12.4 Critique of the N/S/A Distinction: Re-engineering It as a Working Opposition

It is not difficult to think of plausible counterexamples to a sharp dividing line between natural, social, and artificial kind terms: A rock being used as a crude hammering device, the artificial canals of Amsterdam, engineered biological organisms, and genetically modified crops. During the writing of this chapter, headline news broke of the creation (if creation is indeed the right word) of a 'synthetic' cell.[5] Lynne Rudder Baker cites digital organisms, robo-rats, bacterial batteries, and 'search-and-destroy' viruses as examples that test a boundary between natural and artificial kinds (2008). It is clear that if there is a divide at all, it is not a clear one. This blurriness is likely a condition of all non-logical oppositions. It is important to remember that, although such tensions and imperfections may exist, the distinctions and boundaries we invent serve or have served a purpose. Further, if there is nothing new about these problematic cases, there is less reason to suppose that a once useful distinction has fallen into desuetude. Most entities, states, events, and processes do not pose a problem for the N/S/A framework, and although some challenges may prove fatal to the framework in the future, something will be lost in terms of heuristic value.

The way in which I propose we look at such distinctions is not as logical-binary oppositions but as *working oppositions*. CAT argues that all classifications should be interpreted with reference to the sociotechnical context in which they were created. That is, with reference to the particular social and technological systems that sought to address a particular need to classify objects in a particular way. One cannot conceive of something and use it as a microscope, for example, without being clear on the natural objects in one's environment and the social context that are relevant to the investigation.[6] The activity of making distinctions is not meant to preclude counterexamples. It is meant to be a useful way to split up referents that are spoken about in markedly different terms and which it is necessary to treat in markedly different ways. We should not expect all references to correspond neatly to natural, social, or artificial kind terms. Instead we should aim to identify regularities and patterns in classificatory practices which can inform and be informed by

---

[5] *The New York Times*, May 21, 2010, p. A17.

[6] I am not denying that often microscope users have not, before the investigation, known each and every natural entity that they will use the microscope to see. Discoveries have been made with microscopes, of course. Even so, it remains the case that one must have a general conception of what things in one's environment are natural and what things are artificial.

philosophical theorizing. Working oppositions do not comprise an airtight ontology but a pragmatic solution to picking objects out of the world and grouping them into useful categories. CAT also holds that we should be reflexive about these commitments. Philosophical ontology should be analysed in the same terms as classificatory practices in the natural sciences and engineering (Cf. Bloor 1976, pp. 173–174; Hesse 1974).

Given that practices of explicitly classifying objects are familiar to many professional engineers, the question arises how the CAT classificatory practice relates to these other practices. Many companies today employ 'engineering ontologists' who are given the specific task of constructing an often quite blunt system for organizing and ranking objects such as tools within the company. These are often arranged according to a type and given serial numbers or names which reflect a designed hierarchy. Similarly many company manuals and reference documents will contain ordered tables of, for instance, what each tool is called and what it does. The collectivist account hopes to acquire a better understanding of classifications within these communities by looking for general principles and patterns of classification which may hold across a range of organizations and professions. The Kusch-Barnes approach to kinds is perhaps more esoteric than the methods used by engineering companies but has a number of virtues. For one thing, it is adaptable to variances between communities and adaptable over time without allowing mere spontaneous adaptation of proper application (except in highly atypical cases). If Marco Polo used the term 'Madagascar' to refer to the island and not, as was apparently normal then, part of the African mainland, then he would likely have been ignored or corrected by his contemporaries.[7] This single out-of-place use would not change the proper application of the term. However, over time, others did begin to adopt this application and so the reference changed. One vice that may be lodged against the account is that by focusing on referencing practices, it is overly superficial, whereas traditional metaphysicians have looked for something more robust and fundamental. Perhaps Kusch or Barnes would rejoin that this search is wrong-headed but, even if it is not, the accusation of superficiality is, I submit, unjustified. For instance, a traditional realist or essentialist metaphysician might state, as Hilary Putnam and Saul Kripke did, that the 'essence' of water is the microstructure that we now classify as H_2O (Kripke 1980; Putnam 1975). That is, the true referent is not fixed by identifying a string of empirical characteristics but by the molecular composition of water, H_2O. However, there seems to be no principled distinction here. We might ask how the molecular composition is identified: it is through various tests, all of which are empirical investigations. There is no principled method for distinguishing between 'is wet', 'flows in lakes', and 'is tasteless, odourless, and transparent' and passing a scientific test for having the molecular composition H_2O or, as is common

[7] The example of Madagascar as a case of reference switching is taken from Evans (1973). Evans' suggestion is that when Marco Polo used the term to refer to the island we now call Madagascar, he intended to use it as it had been used by others, not introduce a new usage. The lesson I illustrate here is that once Polo did use it to refer to the island, and this stuck, the proper usage subsequently switches.

for identifying water in a scientific setting, testing neutral under a pH test, or boiling at 100 °C at sea level. In each of these cases, proper application of the reference is determined by comparing a set of empirical characteristics and behaviours with previous exemplars. Methodologically speaking, the analysis of lay and scientific reference can be the same. Thus, the collectivist account can refute the essentialist's rejoinder by pointing to the fact that essentialists have not provided a systematic method for getting at the essence of something without including a model akin to Barnes' pattern recognition.

12.5 Applying the Collectivist N/S/A Framework to a Case Study

12.5.1 Textual Analysis of Referencing Practices

The N/S/A distinction rests on there being distinctive referencing practices about natural, social, and artificial kinds. That there are natural and social kinds I take, at least for our purposes here, not to be in dispute. Here I discuss only CAT, which relies on the existence of such practices for artificial kinds. Although for some readers this claim will seem to be justified on the basis of the previous discussion, I take it that CAT requires empirical justification if it is to inform philosophical ontology. The following case study is drawn from the world of petroleum engineering. A world, we might say, in which artificial things extract natural things from the Earth in order to meet needs and conditions determined by social things.[8] Five documents were selected from four interrelated disciplines. Wellbore navigation, directional drilling, and gyroscopic surveying are some of the specialist fields involved in drilling an oil and gas well towards a predetermined target. Production logging provides techniques and heuristics for optimizing production by measuring various properties of fluids in the well. In short, these disciplines use tools to measure properties of a well (be it the orientation and inclination of the well or the properties of fluids such as velocity, density, and temperature). The term 'tool' was chosen as the term to study for this project due to its frequency and relevance.[9]

[8] That is, a world in which oil rigs, wells, tools, and many other artificial things are used to extract hydrocarbons (oil, water, and gas) from rock formations below sea level according to the needs and requirements of society. Similarly, the distinction is present in the children's game 'rock, paper, scissors' from which I borrowed the title of this paper although it requires some imagination on the part of the reader as 'rock' is not a typical candidate for natural kind and paper ought to be read as writing, document, certificate, money, or the like (and thus a social kind).

[9] There were a total of 1708 references to 'tool' or 'tools' in a total of 914 pages or 1.87 references per page. Compare this frequency with the term 'oil' which one might expect to be a popular topic in petroleum engineering but in fact appears only 51 times in four of the selected documents. This equates to 0.09 references per page.

The documents break down as follows: two generic textbooks, one on well logging (Darling 2005) and one on production logging (Scientific Drilling International, Inc. 2010); one handbook for surveyors and directional drillers at BP Amoco (1999); and two operations manuals for specific tools used by Scientific Drilling International, Inc. (Kaulback 2009; Hawkinson 2000). Each reference to 'tool' was selected and defined according to what else was mentioned in that sentence. In consultation with engineers and production logging analysts, five possible types of reference were chosen: informational, proper function, reliability, societal, and other. 'Informational' was broken down into three further categories: measurement, transmission, and interpretation. Each of these three was assumed to be subsets of what may be termed informational reference in that they referred to the capacity of a tool to provide information about the natural (and sometimes artificial or social) world.

Category I. 'Measurement' mentions are those which refer to the capacity of the tool to measure some property. 'Transmission' mentions refer to how, why, in what form or for what purpose (and so on) the tool transmits its information. 'Interpretation' mentions include those which describe how an analyst or engineer receives, interprets, analyses, and uses this information, formulae used, and so on. For example,[10]

> The Dresser Atlas Spectralog tool measures the counting rates in a number of "windows", each of which spans a certain energy band. (Serra 1984, p. 114)

This would be counted twice: once as a tool reference to what the tool *measures* and once to the proper function (see below) of the tool. The following is an example of transmission:

> Tool will take a set of data, and display output on screen. (Hawkinson 2000, p. 16)

Category II. 'Proper function' is a more straightforward category relating to what the tool is *for* and how to use it in order to achieve that goal. We also decided to include in these references to how the tool works, references to the structure, calibration, or alternative features of the tool. For example,[11]

> The Teledrift tool comprises of a pendulum that moves along a series of graduated stop shoulders and a signaling plunger at the top that traverses a series of annular restrictions to produce pressure pulses in the mud flow. (Scientific Drilling International, Inc. 2009, p. 6)

An example of a reference to an alternative calibration or feature would be

> A second mode of operation, known as High Speed, allows the tool to survey on-the-fly once inclination reaches 15 degrees or more. (Hawkinson 2000, p. 6)

Category III. 'Reliability' is, perhaps surprisingly to laypeople and of particular interest to reliabilist epistemologists (possibly the most common breed in epistemology),

[10] Serra (1984) is not one of the texts analysed.

[11] Scientific Drilling International, Inc. (2009) is not one of the texts analysed.

a frequent and persistent concern on a par with the previous types of reference. We include here a broad definition of reliability comprising statistical and probabilistic uncertainty in information provided by the tool as well as the advantages and disadvantages of employing particular tools over others in different geographical locations, rock formations, and situations. The following would be an example of a tool reference to advantages a tool has over another and to the uncertainty or error possibilities in the measurement:

> The Schlumberger Natural Gamma ray Spectrometry tool uses five windows, making fuller use of the information in the spectrum so as to reduce the statistical uncertainty on the analysis of Th, U and K [thorium, uranium and potassium]. (Serra 1984, p. 114)

An example of advantages/disadvantages may include references to current recommendations or histories of the tool:

> …[T]he FDP [fluid density differential pressure] is maintenance intensive and involves the use of mercury, so this tool is being phased out & will be replaced the fluid density delta-p (FDD) tool. (*sic*) (Scientific Drilling International, Inc. 2010, p. 10)

Category VI. Whilst accepting that from a CAT perspective, it is methodologically unsound to disentangle social interests from ontological and epistemological claims, we reserved a category for what we have called 'societal' references. This is taken from a plausible insight into the petroleum engineering world that certain issues are isolated and often precedent to other concerns. Safety and risk are two such issues which are given high prominence in the education, instruction, and day-to-day life of engineers in this field. This is unsurprising especially when one considers the catastrophic, sometimes fatal, effects of accidents, as witnessed in the recent Deepwater Horizon blowout.[12] Reference to hiring costs of equipment, the organization and hierarchy of personnel, who to speak to in order to obtain a particular tool, and so on, were also termed 'societal'. For example,

> All non essential personnel will not be allowed to be in the direct vicinity of the source as it is being loaded into the tool. (Kaulback 2009, p. 130)
>
> As can be seen from the diagram below the "hottest" point at the surface of the tool gives 2.10 μSv/h so it would not be possible to reach anything like a classified workers exposure limit in a week! (Scientific Drilling International, Inc. 2010, p. 15)

Category V. 'Other' references are those which did not easily fit into any of the previous four categories. This category also includes references to naming conventions, other documents, etc. which were thought not to be indicative of the ontological status of an entity. References that purely served the format of a text (e.g. in indexes or tables of contents) and those in discussions of what follows or of the layout of the chapter were excluded.

---

[12] A timeline of which incident is currently available online at http://www.offshore-technology.com/features/feature84446/. 2010. Deepwater Horizon: A Timeline of Events. *Offshore Technology* (Net Resources International).

It was not known before analysis how frequency rates for the different types of reference would be split and consequently what features, qualities, patterns, norms, and so on, were given prominence in engineering classification.

12.5.2 Analysis of Results

The word 'tool' features throughout these texts as a catch-all term for technical artifacts which exclude semi-permanent artifacts such as rigs and pipes. In total, 1708 tool references were found in the five texts.[13] The results revealed one part of the big picture of how production loggers and directional drillers (and surveyors, wellbore navigators, etc.) refer to tools. It was noted that technical operations manuals are likely to contain many mentions of the word 'tool' because it will contain repeat and precise instructions for what to do with the tool and each of its configurations or variants. One textbook (Kaulback 2009) was unusual among the selection in having almost 60 % of its total tool references refer to how to use the tool or how the tool works – in other words, the proper function of the tool. The other four textbooks displayed a statistically even spread across three types of reference: informational, functional, and reliability. 3–11 % of references to a 'tool' were of societal or other type. If we exclude the operations manual mentioned above (which I suggest skews the significance of each reference by repeating very similar mentions), 32–38 % of references were informational and 22–31 % were to proper function. Finally, 24–37 % of tool references have to do with the uncertainty, error, or 'corruption' of the data provided by instruments. The average proportion in that case calculates as follows: approximately one third (34 %) of tool references to what the tool measures/communicates or how these communications are interpreted, approximately one third (32 %) to the uncertainty of data, approximately one quarter (26 %) to proper function, and the remainder (8 %) were to 'societal' and other types of reference.

The large number of references which fitted into three general types suggests that those interested in the ontology of technical artifacts ought to look to the following definition when considering the nature of measurement tools in petroleum engineering:

> (T1) Measurement tools are functional instruments that measure certain properties and provide information to a receiver within a certain range of accuracy.

These three elements seem indispensable for describing the kind of 'measurement tool' spoken of in the documents to a sufficient degree of detail. So far this case study does not, in itself, support a collectivist account. In fact, all case studies such as this are able to do is build up a body of research which can be given a collectivist

---

[13] The results are compiled in Table 12.1 in the Appendix.

analysis and from which analysis fruitful insights may be drawn. Future research may be able to generalize the results of this study or perhaps adapt the criteria in order to make it more generalizable.

12.6 The Collectivist Aspect of Tools

One common counterexample to collectivist approaches such as CAT is a variation on a theme which I will call 'Robinson Crusoe' examples. The relevant variety here is 'Robinson Crusoe's tool'. Consider the following remark from Amie Thomasson:

> [I]t makes perfect sense to suppose that a solitary human could create a knife, but not a government or money. Thus artefacts don't seem to be essentially *social* objects at all. (Thomasson 2007, p. 52. Original emphasis)

Suppose Thomasson's solitary human was trying to create the first ever knife. Is it still intuitive to say that this person can do that but not create the first ever government or money? It seems that Thomasson's statement only makes sense if one first supposes that there exists a community of tool users (knife users, scissor users, hammer users, etc.) and that this solitary human is creating a tool of a kind which is recognized and established within the community; a kind of which it makes sense to say that it is an artifact only within the system that defines what counts as an artifact.

> **Crusoe's Tool**
> Robinson Crusoe is a man stranded on a desert island. He was stranded there as a young child and all he has ever known are natural objects – trees, plants, fruit, sand, and so on. Nevertheless, he makes use of what he has around him to fashion a rudimentary axe, knife and various other useful things. He uses them to accomplish various tasks, and makes other similar objects, gradually improving them each time.

Has Crusoe created tools? Some readers will say, 'Yes, of course he has created tools. As you say, he has created axes and knives. What are these if they are not tools?' Note that I said he has created a 'rudimentary axe'. Alternatively, I might say that he has created a proto-axe or an axlike object. This would minimize any preconception about whether or not he has created a tool. Now let us consider what it would mean to say that he has created a tool. Would we also have to say that you can create a tool even if no one else recognizes it as a tool? Perhaps we can say that they *would* recognize it were they to see it. But CAT holds that one cannot recognize a tool just by looking at it. One must also know the communal structure that makes it a tool. That seems to be missing here since only Crusoe knows of the object's existence. Beth Preston has argued that tools are typically identifiable regardless of our witnessing anyone using the tool:

> [U]nlike naturally occurring objects, manufactured tools are designed for a specific use, and normally have a standardized form. More importantly, they are identifiable as tools—indeed often as specific kinds of tools apart from *any* actual occasions of use. (Preston 1998, p. 6. Original emphasis)

Whilst this may often be the case, it seems anthropologically flawed. The only reason why we would ever be able to identify what a tool is for would be on the basis of a) witnessing actual use and b) prior knowledge as to the proper function of similar tools (Vaesen and van Amerongen 2008; Dennett 1990). However, this serves to emphasize the need for a collectivist analysis. We need to first place the behaviour within some communal structure in order to classify the tool, usually by observing how Crusoe uses the tool and either assimilating that into an already existing kind of tool use or, less commonly, by constructing with the user a communal structure for using that kind of tool. Consider a variation on Crusoe's tool:

> **Crusoe and the Coconut**
> Robinson Crusoe is lonely on the island and decides to create a friend. He draws a face onto a coconut, places it on a stick, and begins to talk to it. Since Crusoc has been on the island since he was a young child he has never learnt to speak English and so he makes up his own noises to talk to (or, rather, at, for it cannot respond) the coconut.

The scenario is the same as Crusoe's tool, although in this case the question is over whether Crusoe has created a language rather than a tool. I think in this case we are unlikely to say that Crusoe has created a language, especially those of us who are familiar with Wittgenstein's later work. As Wittgenstein said in his *Philosophical Investigations*,

> The words of this [private] language are to refer to what can be known only to the speaker; to his immediate, private, sensations. So another cannot understand the language. (Wittgenstein 1953, para. 243)

Just as Crusoe cannot create a language by himself (since a language requires communication and community), he cannot create a tool by himself (since tools require proper functions and community). Crusoe may construct something that looks like a tool, functions as a tool, and is intended to fulfil a specific function. Crusoe cases are rare but they highlight a specific issue that is underplayed in Thomasson's and Preston's account. We require a social background of tool use in order to know what the proper function of a particular kind of tool is and, once established, that background knowledge can be extremely powerful. However, none of that implies that we can separate our analysis of what the proper function of a tool is from the actual occasions of use. The individual causal history of the artifact and the nonindividual normal use history, to borrow Preston's phraseology, do not come apart but are both vital to any understanding of that particular tool and its function within a community. This point is more evident when we consider not simple or basic tools such as hammers and screwdrivers but much more complex technological artifacts such as aeroplane engines or Hadron Colliders, since the background knowledge of tools of that kind may be quite limited and weak (since they are fewer and their occurrence is limited to specialized zones). Crusoe's makeshift tool cannot be properly referred to as a tool because it exists in isolation from a community and is consequently without proper function.

A similar conclusion is reached when we look at tool appropriation. Tool appropriation occurs when an object (either purely physical or artificial) is used *as* a tool although it is not itself a tool, or not that kind of tool. For example, suppose that

I am unfamiliar with surgeon's scalpels and, on finding one in a colleague's desk drawer, I use it as a letter opener. In so doing, it seems we should say that I have not instantly changed the scalpel into a letter opener but am using a scalpel *as* a letter opener. In other terms, I am taking the scalpel to be a letter opener, referring to it as such. But this is just one instance among a whole classificatory system, and, what is more, I have no particular authority in the community when it comes to deciding what surgeon's scalpels are for. Consider instead a more mundane example where you walk into the forest and construct a rudimentary lever using a branch. The collectivist account analyses this as using a branch *as* a lever, not as creating a lever (since it is an individual action). This is an extension of the principle discussed earlier that technical artifacts must be purposefully designed for a particular function or functions. It is not enough to pick up a palm leaf and use it to fan yourself for that palm leaf to stop being a natural kind and become a technical artifact (i.e. an artificial kind). For this to be the case, social structures of proper use would have to be implausibly flexible and transitory. Secondly, the function (of fanning) is incidental to its physical structure; it was not purposefully designed for this function. Further, referring back again to the five criteria outlined previously, there are no normative constraints on the palm leaf. No one is saying that the palm leaf *ought* to be used to fan. There might be occasions where someone would do this but the whole point of appropriation is that there is not this complex normative structure already in place. It is a fleeting moment of innovation. If, on the other hand, there exists a community who routinely appropriates palm leaves for the function of fanning, who amends their physical characteristics (or, at least, executes some design decision upon them), and who collectively says 'You should use palm leaves for fanning', then the cut palm leaves may change their ontological status from natural to artificial kind, from simple leaf to tool. There is undoubtedly a grey area concerning this transition, and whether it is right to call the object natural or artificial is a contingent, not an absolute, matter. Any analysis of the transition has to pay close attention to the reference talk of the community, its design choices, and the development and use of the putative tool.

12.7 Conclusion

In this chapter, I presented an application of the methodology of the Collectivist Account of Technical Artifacts to a particular engineering case. I also discussed some of the theoretical issues around CAT and staked an empirical and philosophical claim for the ontology of technical artifacts. I began by submitting that philosophical discussions about the ontology of technical artifacts ought to be informed by the ontological assumptions underlying engineering practices. The vast literature attempting to define the ontological assumptions underlying scientific practices may have to be emulated for the engineering sciences and technology. To aid in this, I provided a particular conceptual framework for deriving these practical ontologies, namely, CAT. Following Martin Kusch, I suggested that an intuitive ontology conceives three kinds or kind terms: natural, social, and artificial. CAT aims to build on this ontology and provide an empirical way of analysing the nature of the three kind terms. CAT views the differences in kind terms as differences in reference talk. If we analyse the reference talk of a community, we can distinguish different kind terms as well as observe how the three are conceptualized and construct themselves. I describe how Kusch's account distinguishes the three kind terms by virtue of reference talk taking them to be such and such a kind and how CAT can be used to analyse actual instances of reference talk.

I have added an example of CAT in action and suggested that this project should be expanded upon in future research in order to validate the results found. The example I presented analyses the textual output of two branches of petroleum engineering. I analysed the kinds of reference made to 'tools' – a kind of technical artifact – and separated these references into seven categories: measurement, transmission, interpretation, reliability, function, societal, and other. It turns out that these categories can be said to be dominated by three kinds of reference talk: informational, reliability, and function. These three categories – the first encompassing the measurement, transmission, and interpretation categories – are evenly split in the number of their mentions in the texts. It was noted that such categories do not apply easily to natural and social kind terms in general. This indicates that these three categories are crucial to the ontological assumptions underlying engineering practice within this community. If further research corroborated these results, we may be in a position to say that this should also inform a philosophical ontology of technical artifacts.

Acknowledgments The fieldwork research for this study was conducted primarily at the offices of Scientific Drilling International, Inc. in Thailand and other parts of Southeast Asia as part of my PhD dissertation. I would like to thank everyone there who let me observe their day-to-day work, interview them, and who instructed me in the basics of their work. I would also like to thank engineers at Chevron, Halliburton, and Schlumberger for allowing me to interview them and for their feedback on my project and development of the five referencing categories. I thank Maarten Franssen, Peter Kroes, and Pieter Vermaas for the invitation to contribute a chapter to this volume and for their astute comments and criticisms along the way. I would also like to thank David Bloor, Martin Kusch, Duncan Pritchard, and Pablo Schyfter for their insightful feedback and conversations on this project.

Appendix

Table 12.1 'Tool reference' in five petroleum engineering texts

Source	Mea	Tra	Int	Inf	Rel	Fun	Soc	Oth	Tot
BP Amoco (1999)	64	23	41	128	149	102	21	2	402
%				32	37	25	5	1	100
Hawkinson (2000)	60	28	20	108	98	99	10	3	318
%				34	31	31	3	1	100
Darling (2005)	43	7	51	101	83	61	30	5	280

(continued)

Table 12.1 (continued)

Source	Mea	Tra	Int	Inf	Rel	Fun	Soc	Oth	Tot
%				36	30	22	11	2	100
Kaulback (2009)	33	36	12	81	117	364	58	8	628
%				13	19	58	10	1	100
SDI (2010)	22	2	6	30	19	20	9	2	80
%				38	24	25	11	3	100
Total	222	96	130	448	466	646	128	20	1708
Total %				26	27	38	7	1	100

When Kaulback (2009) is excluded, the total percentages become, respectively, 34, 32, 26, 1, and 6
The number of pages in each document in the order listed in the table are 278, 91, 355, 175, and 15
Key to Table 12.1
Source means the text used
Mea means number of tool references to measurement
Tra means number of tool references to transmission
Int means number of tool references to interpretation
Inf means number of tool reference to informational (the combined total of tool references to measurement, transmission, and interpretation)
Rel means number of tool references to reliability
Fun means number of tool references to proper function
Soc means number of tool references to societal
Oth means number of other tool references
Tot means the total number of references and percentages thereof

References

Austin, J. L. (1975). *How to do things with words*. Cambridge, MA: Harvard University Press.

Baker, L. R. (2004). The ontology of artifacts. *Philosophical Explorations, 7*, 99–112.

Baker, L. R. (2008). The shrinking difference between artifacts and natural objects. *APA Newsletter on Philosophy and Computers, 7*(2), 2–5.

Barnes, B. (1983). Social life as bootstrapped induction. *Sociology, 17*(4), 524–545.

Bird, A., & Tobin, E. (2008). Natural kinds. In: *The Stanford encyclopedia of philosophy* (Autumn 2008 ed). http://plato.stanford.edu/entries/natural-kinds/

Bloor, D. (1976). *Knowledge and social imagery*. London: Routledge & Kegan Paul.

Bloor, D. (2007). Epistemic grace: Antirelativism as theology in disguise. *Common Knowledge, 13*, 250–280.

BP Amoco Upstream Technology Group. (1999). *Directional survey handbook*. Aberdeen: ODL.

Bunge, M. (1977). *Treatise on basic philosophy: Ontology I: The furniture of the world* (Vol. 3). Dordrecht: Springer.

Darling, T. (2005). *Well logging and formation evaluation*. Amsterdam: Elsevier.

de Koninck, C. (2008). Reflections on the problem of indeterminism. In R. McInerny (Ed.), *The writings of Charles de Koninck* (Vol. 1, pp. 401–442). Notre Dame: University of Notre Dame Press.

Dennett, D. C. (1990). The interpretation of texts, people and other artifacts. *Philosophy and Phenomenological Research, 50*(S), 177–194. (Reprinted in M. Losonsky (Ed.) *Language and mind: Contemporary readings in philosophy and cognitive science*. Blackwells.)

Dupré, J. (1993). *The disorder of things: Metaphysical foundations of the disunity of science*. Cambridge, MA: Harvard University Press.

Evans, G. (1973). The causal theory of names. *Proceedings of the Aristotelian Society, Supplementary Volume, 47*, 187–208.

Garbacz, P. (2012). What can an armchair philosopher do for a "dirty-hands" engineer. *Axiomathes, 22*(3), 385–401.
Hacking, I. (1991). A tradition of natural kinds. *Philosophical Studies, 61*, 109–126.
Hawkinson, B. (2000). *Keeper operation manual.* (Reprinted 2008 as *AdK, Casing multishot, Drillpipe multishot, Memory slickline, single shot orientation*. Paso Robles: Scientific Drilling International Inc.)
Hesse, M. E. (1974). *The structure of scientific inference*. London: Macmillan.
Hoffman, J., & Rosenkrantz, G. S. (1997). *Substance: Its nature and existence*. London: Routledge.
Kaulback, M. (2009). *Memory production logging: Operations manual V1.6*. Paso Robles: Scientific Drilling International Inc.
Kripke, S. (1980). *Naming and necessity*. Cambridge, MA: Harvard University Press.
Kusch, M. (1997). The sociophilosophy of folk psychology. *Studies in History and Philosophy of Science, 28*, 1–25.
Kusch, M. (1999). *Psychological knowledge: A social history and philosophy*. Oxford: Routledge.
Layton, E. T., Jr. (1974). Technology as knowledge. *Technology and Culture, 15*(1), 31–41.
MacIntyre, A. (1984). *After virtue: A study in moral theory* (2nd ed.). Notre Dame: University of Notre Dame Press.
Okasha, S. (2002). Darwinian metaphysics: Species and the question of essentialism. *Synthese, 131*, 191–213.
Pitt, J. C. (2011). Working the natural/artificial distinction. *Philosophy of Engineering and Technology, 3*(2), 73–83.
Preston, B. (1998). Cognition and tool use. *Mind and Language, 13*(4), 513–547.
Preston, B. (2008). The shrinkage factor: Comment on Lynne Rudder Baker's "The shrinking difference between artifacts and natural objects". *APA Newsletter on Philosophy and Computers, 8*(1), 26–28.
Putnam, H. (1975). The meaning of 'meaning'. In *Mind, language and reality: Philosophical papers* (Vol. 2). Cambridge: Cambridge University Press.
Schyfter, P. (2009). The bootstrapped artefact: A collectivist account of technological ontology, functions, and normativity. *Studies in History and Philosophy of Science, 40*, 102–111.
Scientific Drilling International, Inc. (2009). *Basic survey manual*. Paso Robles: Scientific Dilling International, Inc.
Scientific Drilling International, Inc. (2010). *Production logging*. Paso Robles: Scientific Drilling International Inc.
Searle, J. (2002). What I am not a property dualist. *Journal of Consciousness Studies, 9*(12), 57–64.
Serra, O. (1984). *Fundamentals of well-log interpretation*. Amsterdam: Elsevier.
Sober, E. (1994). Evolution, population thinking and essentialism. In E. Sober (Ed.), *Conceptual issues in evolutionary biology* (2nd ed., pp. 161–189). Cambridge, MA: MIT Press.
Thomasson, A. L. (2007). Artifacts and human concepts. In S. Laurence & E. Margolis (Eds.), *Creations of the mind: Essays on artifacts and their representations* (pp. 52–73). Oxford: Oxford University Press.
Vaesen, K., & van Amerongen, M. (2008). Optimality vs. intent: Limitations of Dennett's artefact hermeneutics. *Philosophical Psychology, 21*(6), 779–797.
Vincenti, W. G. (1990). *What engineers know and how they know it: Analytical studies from aeronautical history*. Baltimore: John Hopkins UP.
Whitrow, G. J. (1980). *The natural philosophy of time* (2nd ed.). Oxford: Oxford University Press.
Wittgenstein, L. (1953). *Philosophical investigations*. Oxford: Blackwell.

MIX
Papier aus verantwortungsvollen Quellen
Paper from responsible sources
FSC® C105338

If you have any concerns about our products,
you can contact us on
ProductSafety@springernature.com

In case Publisher is established outside the EU,
the EU authorized representative is:
Springer Nature Customer Service Center GmbH
Europaplatz 3, 69115 Heidelberg, Germany

Printed by Libri Plureos GmbH
in Hamburg, Germany